HMS BULWARK 1948–1984

Ken Burns

Mike Critchley

MARITIME BOOKS

1986

Field — Barry Wavy of Ten White and Blue
Badge — A Sword erect proper the blade enfiled with a crown palisade gold

Contents

First published in Great Britain
by Maritime Books, Liskeard, Cornwall

ISBN 0 907771 28 9

Printed in Great Britain by Penwell Ltd.,
Parkwood, Callington, Cornwall

1

Her Construction
1944 to 1954

Bulwark *enters the water for the first time at Belfast 22 June 1948.*

Fitting out at Belfast, 20 June 1953.

The Light Fleet Carrier originated during the 1939-45 war with the demand for more and more aircraft carriers capable of working with the Fleet on offensive operations. The requirements were a good turn of speed, great steaming endurance and facilities for operating and maintaining aircraft over prolonged periods without external assistance. The construction had to be simplified to permit rapid building, so some of the qualities of the Fleet Carrier had to be sacrificed.

The design was the subject of a long series of model experiments to determine the best shape of hull for speed, endurance and seaworthiness. The strength and rigidity of the hull was also the subject of special investigations.

The Hermes Class of Light Fleet Carriers was to comprise *Albion, Arrogant, Bulwark, Centaur, Elephant, Hermes, Monmouth* and *Polyphemus*. Of these eight ships, four of them—*Arrogant, Hermes, Monmouth* and *Polyphemus*— were cancelled. The *Elephant* was renamed *Hermes*, but when completed she was so much different to the other three, that she was virtually in a class of her own. The remaining ships, *Albion, Bulwark* and *Centaur* became enlarged versions of the *Hercules* class with machinery of twice the power to increase their speed by 5 knots.

Bulwark was authorised in the 1943 naval programme and was laid down in the Musgrave Shipyard of Harland and Wolff Ltd, Belfast on 10 May 1945. In the construction of the hull, electric welding was used extensively. X-ray testing of welding had become standard procedure in shipyards and the team of young scientists with their radiographic equipment was a familiar sight to the workmen onboard *Bulwark*.

The geared turbine machinery was made by Messrs Harland and Wolff at their Queen's Island Works, and embodied the latest practice. The increasing use of electric power in warships called for considerable generator power which was provided by both turbine and diesel dynamos, with elaborate safeguards for the maintenance of supplies to all services in the event of damage. It was estimated that more than 250 miles of electric cable were incorporated into the ship.

During the course of building, a number of changes were made to the design to bring it completely up-to-date. The living quarters of the ship's company were arranged on the same system as had been use in the battleship *Vanguard* with spacious dining halls close to the galleys, and with mess decks equipped with tubular steel furniture. The catering arrangements included a high

Sisters under construction—**Centaur** *(left) and* **Bulwark** *at Belfast, 20 June 1953.*

degree of mechanisation, with the cooking done mainly by electricity. More space than usual was allocated to refrigerated stowage in 'cold' and 'cool' rooms, to ensure a more varied diet than had hitherto been possible in warships.

When the Second World War ended, the pace of building eased, so the ship was not launched until 22 June 1948. The traditional service, prescribed by the Admiralty for the launch of HM Ships, was conducted by the Rev A T I Ford, MA, Rector of St Patrick's Parish Church, Ballymacarrett. The singing was led by the Queen's Island Male Choir accompanied by the Band of the Royal Ulster Constabulary.

At 12.30pm that day, Her Excellency the Countess of Granville, wife of the Governor of Northern Ireland, and sister of H.M. Queen Elizabeth, named and launched *Bulwark*—the ninth aircraft carrier to be launched at Belfast. For a moment after she had pulled the launching lever there was no movement, but then the hull, painted in a new shade of Naval grey, moved down the slipway. As it did so, an RAF flying boat flew low over the shipyard—a striking salute by one Service to another.

There were many other distinguished guests including the Prime Minister of Northern Ireland (Captain the Rt Hon Sir Basil Brooke, Bart) and Lady Brooke; members of the Northern Ireland Cabinet and their wives; Admiral Sir Frederick Dalrymple-Hamilton, Flag Officer Commanding Coast of Scotland and Northern Ireland; and Sir Frederick Rebbeck, Chairman and Managing Director of Harland and Wolff Ltd; and Lady Rebbeck.

At the after-launch luncheon, Sir Frederick Rebbeck presented Countess Granville with a Scottish Silver Rose Bowl which had been made in 1785.

The paint on the hull of the carrier was a new synthetic resin paint with which the Admiralty were experimenting. In addition to saving linseed oil, it was expected to give a better finish, reduce the risk of fire and have an increased durability.

In 1950, and with the outbreak of the Korean War, the pace of work in *Bulwark* was stepped up. Under the command of Captain J M Villiers, OBE, RN, she commissioned on 29 October 1954 for trials preparatory to assuming the role of a Trials and Training Carrier.

2

First Commission

1954 to 1956

Captain Villiers (second from left) signing for 14 million pounds worth of ship, 15 December, 1954.

Early in November 1954, *Bulwark* sailed from Belfast for Portsmouth, where, during the next few weeks, the ship's company were kept busy fitting many items of equipment including much of her flight deck 'extras'. On 15 December she was accepted into service when Captain J M Villiers signed for £14 million pounds worth of ship.

A suitable reward for the first pilot to land onboard—Lt Cdr Nance of the US Navy.

Bulwark's seasonal leave and self-maintenance period alongside in Portsmouth ended on Saturday 29 January 1955 and the ship proceeded to Spithead on the Monday to carry out various calibrations and trials.

Flying trials were due to start on 7 February, but high winds caused a delay as they gave unsuitable conditions for testing the pull out of the arrester gear. Later that day, however, conditions improved and the trials commenced. The first deck landing ever to be made on *Bulwark* by a conventional fixed wing aircraft was carried out by an American! An Avenger from RN Air Station, Ford was piloted by Lieutenant Commander J W Nance, USN, of Raleigh, North Carolina, who was on an exchange appointment to the Service Trials Unit.

Pilots of 807 Squadron on board their new home.

During the next 3 days, proving trials of the catapult gear were carried out without incident with differently loaded Avenger, Sea Hawk and Skyraider aircraft. The arrester gear trials were also conducted by these aircraft with additional sorties flown by Wyvern and Gannet aircraft.

The following week, trials commenced to evaluate the Mirror Deck Landing Sight for the Avenger and Skyraider aircraft. In addition to flying trials, surface exercises were carried out with the destroyer *Obdurate*, including refuelling abeam, transfer by jackstay, and taking, and being taken, in tow.

Ground maintenance personnel for the ship's own aircraft were embarked at Portland on 19 February, and the ship sailed the following day for the Moray Firth. En route, two Skyraiders of 849 D Flight and the Ship's Flight Avengers were embarked, together with nine Seahawks of 807 Squadron, each of which carried out two 'touch-and-go' approaches before 'hooking on'. The ship passed through the Dover Straits in a heavy snow storm and an unpleasant night was spent in the crowded shipping lanes of the southern North Sea.

The ship's work-up period continued with navigation and interception exercises, using the Seahawks of 807 Squadron. Carrier controlled approach trials were carried out to establish the best methods of marshalling and the type of flying pattern to be employed as standard drills at sea. On completion of these trials the ship secured to a buoy at Invergordon, where the Royal Marine Detachment made its first landing. It was to carry out exercise 'Handcart' in the hills of Caithness. When the Royal Marines departed on Friday 25 February, no one envied them their icy bivouacs amongst the snow covered hills—temperatures of 20°F were recorded, even on the Flight Deck at sea. Indeed, on the 26th, the mercury literally dropped out of sight, retreating into the bulb of the thermometer which could only register down to 15°F.

On 4 March the ship went alongside at Rosyth to disembark the Squadron ground maintenance personnel and for a short self-refit and maintenance period. 807 Squadron had achieved a total of 180 flying hours and 189 deck landings in their thirteen days onboard, despite flying being possible on only ten of those days.

On completion of her first Spring programme, which, though shorter than originally planned, nevertheless accomplished 852 arrested deck landings, the ship anchored at Spithead late on the night of Friday 25 March. The following day, after discharging Avgas, she proceeded up harbour to await admission to C Lock for dry docking on Tuesday the 29th.

Bulwark's docking and self maintenance period ended on 11 May. After a couple of days deck landing practice for the Skyraider aircraft of 849C flight, and Sea Venoms of 890 Squadron, who were operated directly from RNAS Yeovilton, the ship moored in Torbay. On Sunday 15 May, *Bulwark* was open to visitors for the first time since commissioning and over 1,200 people braved the rather boisterous weather to gain a closer look at the ship.

During the night flying programme on Wednesday 18th, a Sea Venom of 890 Squadron, piloted by the Squadron CO, Lieutenant Commander A Gordon-Johnson crashed into the sea immediately over the bows of the ship after a mishap on the starboard catapult. Both pilot and observer passed underneath the ship in the aircraft and fortunately did not free themselves to come to the surface until the propellers had passed clear over them. They surfaced within twenty yards of *Bulwark's* life raft and life buoys and were quickly seen in the light of the ship's 20″ signalling projector. Their recovery was carried out by the destroyer *Chaplet* who had been acting as plane-guard. The observer, Lieutenant Carter, was able to return to the ship but Lt Cdr Gordon-Johnson was detained in the RN Hospital at Portland for several days.

On 19 May, rendezvous was made with the Royal Fleet Auxiliary *Olna*, and refuelling was carried out at sea. During this operation *Olna*'s quartermaster was seized with toothache, put the helm the wrong way and made a large alteration of course to starboard which led to the parting of all three hoses. Fortunately the amount of oil sprayed on to the recently painted ship's side was not as much, nor as disastrous, as might have been!

Flying operations continued with deck landing trials for 898 Squadron Seahawks. Skyraiders of 849C Squadron carried out a final period of night flying on 24 May to complete the evaluation of the mirror sight for night deck landings by piston engined aircraft.

On Thursday 26th a BBC radio programme was broadcast 'live' from the ship. For the greater part of the

The BBC invade—(26 May 1955) for Children's Hour. Can you recognise Max Robertson, Cliff Michelmore and Raymond Baxter?

. . . wearing the flag of Admiral of the Fleet Sir John Cunningham off Oslo.

broadcast, visibility was down to 100 yards which precluded launching 807 Squadron strike aircraft. However, Lieutenant Commander Jeyes of 890 Squadron flew a Sea Venom from NAS, Ford and carried out a 'touch and go' under difficult conditions.

At Spithead, *Bulwark* embarked the ground maintenance personnel of 801 and 898 Squadrons plus the combined R M Bands of the Flag Officer Air, Home; HMS *Theseus* and the RM School of Music, Deal. Wearing the Flag of Admiral of the Fleet Sir John Cunningham GCB, MVO, DL, the ship sailed for Oslo at 0830 on Monday 30 May. The purpose of Admiral Cunningham's visit to Norway was to present mementoes from HMS *Devonshire* to HM King Haakon VII, the city of Oslo and later, in the destroyer *Saintes*, to the people of Tromso. The mementoes commemorated the close association between Norway and HMS *Devonshire*, which, wearing the Flag of Admiral Cunningham, had brought King Haakon and members of his Government to the United Kingdom when Germany had invaded Norway in 1940.

A flypast over Oslo was carried out by seventeen Seahawks of 898 and 801 Squadrons. After the flypast, four aircraft, led by the Commanding Officer of 898 Squadron, detached and to the south of the city gave a demonstration of formation aerobatics.

A most successful visit included various official functions. Admiral Cunningham was received in audience by HM King Haakon VII to whom he presented the mahogany backboard from the Captain's cabin of HMS *Devonshire*. It bore the ship's crest and battle honours.

On Friday 3 June, HRH the Crown Prince of Norway

Beating Retreat by Massed Bands of the Royal Marines, Oslo, 1 June, 1955.

HRH Crown Prince Olaf inspecting the Officers and Ship's Company, Oslo, 3 June, 1955.

 A fine sight in Stockholm Harbour. Note that all her guns are cocooned.

visited *Bulwark*, inspected the ship's company, and afterwards made a tour of the ship.

A party for about 550 children was held on board. There was also a cocktail party on the quarter deck, towards the end of which the 400 guests proceeded to the flight deck to witness the combined RM Bands perform a ceremonial 'Beating Retreat'.

The ship was opened to visitors on the Saturday afternoon and over 2,000 people came out in boats to see the ship.

There was a considerable exchange of private entertainment in addition to the official functions. The general atmosphere of the whole visit was one of great hospitality and friendship by the Norwegian people.

Bulwark sailed from Oslo on 5 June and, whilst on passage, successful and valuable exercises were carried out with Thunderjets of the Norwegian Air Force. The next day the ship arrived at Rosyth where a strong easterly wind made conditions most difficult for refuelling from RFA *Black Ranger* and the exchange of a large number of Squadron personnel.

Airborne cyclist—the Ship's Dragonfly provided a stunt for the children's party.

A Royal Guard salutes HM King Gustav VI.

From an anchorage at Cromarty, the ship sailed on Friday 10 June for Stockholm. Flying was carried out whilst on passage, and once in the Baltic Sea special interest was added by the presence of Russian aircraft. These took up the roll of 'Shadowers'. A Seahawk aircraft also reported having had a fleeting glimpse of a periscope not far ahead of the ship.

Rendezvous was made with other units of the Home Fleet at 0900 on Tuesday the 14th and the fleet proceeded in company to Stockholm, *Bulwark* and the submarine *Artemis* forming Group 2 to enter harbour. A flypast by 24 Saab aircraft of the Royal Swedish Air Force welcomed the British visitors.

On Friday the 17th HM King Gustav VI of Sweden, attended by Admiral Ericson, the Commander-in-Chief, Royal Swedish Navy, visited *Bulwark*, and inspected the ship's company, together with representative contingents from all the ships in company, at Divisions on the flight deck. On conclusion, His Majesty spent a considerable time visiting the Island hangars, mess decks and various other 'between deck' spaces.

Over 12,000 Swedes looked over the ship during the two days she was open to visitors. In addition, many conducted tours of the ship were given to VIPs.

A small lunch party for the Ship's Company ashore at Stockholm—there was no shortage of volunteers.

The ship benefitted by the presence of the Band of the Highland Light Infantry, which played at all official functions; gave concerts on shore and broadcast over the Stockholm Radio Service. This Band joined with the RM Band and drill squad to perform at the sports stadium after the soccer match in which the British Fleet defeated the Swedish Fleet by four goals to two. The British team included six ratings from *Bulwark*.

During the return passage from Stockholm, a full power trial was carried out to check the temperature of the hot gases leaving the funnel. This trial resulted from reports from some ships that radio equipment had been damaged by excessive heat from the funnel.

Bulwark lay alongside at Portsmouth for 10 days self-maintenance, before sailing on 4 July for the Portsmouth Exercise Area to take on Sea Venoms of 809 and 891 Squadrons. Preparations were then made for three television broadcasts which were to be relayed from the ship. The first of these was transmitted on Wednesday 6 July when television history was made by *Bulwark*. She became the first RN ship to be featured in a live television broadcast. This programme showed the ways in which aircraft were launched and recovered, and included scenes of the arrester gear, mirror sight, catapults, lifts, etc. The "guest star" of the show, though, was a DH 110 which carried out a flypast and a 'touch and go' approach.

Two rehearsals were necessary for the second show which was filmed on the following day. This programme opened with the rescue of a commentator by helicopter net, followed by the launching of Seahawks of 898 Squadron. A strike on the destroyer *Saintes* was carried out by six Seahawks of 898 Squadron augmented by six Seahawks of 800 Squadron, which were operating from NAS *Ford*. *Saintes* was stationed close to *Bulwark* but the cameras had difficulty holding the aircraft during the attacks.

On Friday the ship returned to harbour and preparations began for televising the Garrison Theatre Show to be held on board.

Further flying trials were carried out, followed on Tuesday the 19th by heeling trials. These were carried out at anchor, and a maximum heel of 15 degrees was achieved. The ship weighed and proceeded to continue the trial under way whilst at this angle.

The following Monday the ship went into Portsmouth for the summer leave period and to prepare for Navy Days.

On 15 September she left Portsmouth, and at Spithead embarked six Members of Parliament who were to be given a few days' sea experience while the ship was on passage to Invergordon. After a two day passage, the ship joined the Home Fleet. As soon as she was at Invergordon, 845 Squadron, which was composed of Whirlwind helicopters, transferred from HMS *Eagle*. The ships were anchored close to each other and most of the Squadron equipment was ferried by the aircraft.

Eagle, Bulwark, Centaur *and* **Albion** *at Invergordon.*

Exercise "Sea Enterprise" began on 21 September. This was the first time *Bulwark* had taken part in a major exercise involving an aircraft carrier squadron. The start of flying operations that day was eventful as, within half an hour of the first launch, a "Mayday" message was received from one of the helicopters. This aircraft, which was being flown by the CO of 845 Squadron, ditched a short distance from the destroyer screen positioned ahead of the force. Fortunately other helicopters recovered all members of the crew, including the CO (designate) of the Squadron who was also flying in the aircraft.

The exercise was concentrated on the anti-submarine task, although air strikes took place on the Trondelag area of Norway, and a search, shadowing and attack phase was also carried out on the "enemy" forces.

A fatal accident marred the exercise, when a Wyvern aircraft developed a fuel leak and returned to the force desperately short of fuel. In attempting an emergency landing on *Bulwark*, the aircraft crashed into the deck park and then went into the sea. There was another ditching when the pilot of a Dragonfly helicopter, visiting *Bulwark* from *Eagle*, lost control when the ship rolled heavily just at the moment of take-off. Fortunately all occupants of this aircraft were quickly recovered by helicopters from *Bulwark* and *Albion*.

During the first week of October, deck landing practice was given to three Gannet Squadrons, 825 of the RN and 816 and 817 of the RAN. A total of 347 'touch and go's' and 15 arrested landings was achieved. There were a large number of visitors onboard at this time, including officers of the RAN who were about to be appointed to HMAS *Melbourne*, together with some RNVR officers undergoing training.

The ship anchored at Spithead on Thursday 6 October and next day entered harbour for a week's self-maintenance. During this time the ship was host to the Soviet cruiser *Sverdlov*, the Flagship of a visiting Russian Squadron. Visits were exchanged by ships' officers and on the Saturday, a team of Russian dancers performed in the hangar before the ship's company. The Russians presented a large decorated urn to *Bulwark* as a memento of their stay.

A preliminary trial was also carried out to test the amount of flexing to which the hangar deck would be subjected with the new and heavier aircraft planned to come into service. It is not believed that the Russian dancers were part of this trial.

Bulwark sailed on Tuesday the 18th for the Portsmouth Exercise Areas to embark sixteen Avengers of Nos 2 and 4 Squadrons, Royal Netherlands Navy. Two days later the ship reached Rosyth and joined units of the Home Fleet who were preparing for exercise "Phoenix One". This exercise, which started on the 24th, was designed to test

Heeling trials, south of the Isle of Wight.

the ability of carrier based aircraft to deal with high altitude bombers. It involved using two radar and air direction pickets deployed between 60 and 110 miles from the carrier force. *Bulwark* and the cruiser *Glasgow* fulfilled the role of pickets while the carriers *Eagle, Albion* and *Centaur* formed the main body which was escorted by destroyers.

On completion of the exercise, in heavy weather the ship refuelled using the astern method from RFA *Olna.* A weekend was spent at Rosyth, before sailing for the Moray Firth and further flying operations.

On the first anniversary of *Bulwark*'s commissioning, on 29 October, records showed the ship had spent 188 days, wholly or partly, at sea during which she had steamed 19,272 miles. In that year no less than 55,000 people had visited the ship.

During the ship's passage south to Portsmouth from Rosyth, a start was made on the next main task which was the evaluation of two types of communication equipment. A number of scientists from the Admiralty Signal Research Establishment and representatives of the Standard Telephone Company had joined the ship to assist. The evaluation was carried out in conjunction with UHF fitted Neptune aircraft of the United States Navy operating from Blackbushe. These trials were continued off the Isle of Wight from 7-11 November.

Concurrently a trial was carried out with jet blast deflector plates which had been developed by the Naval Air Division, Bedford. The idea was to test the effectiveness of the deflectors as a protective shield for personnel who would be subjected to the jet blast of engines that were more powerful than those currently in service with the Fleet Air Arm. The fuselage of a Vickers 508 was fitted with two Rolls-Royce Avon engines and was used to produce a jet blast directed at the deflectors which had been welded onto the flight deck.

Bulwark, with aircraft of 700 and 764 Squadrons embarked, visited Cherbourg from 16-21 November. Whilst there, the opportunity was taken to land a small group of Royal Marines who were to visit the landing beaches in Normandy and follow the route taken by 45 Commando RM's on D-Day and D-Day + 1.

On Monday 28 November deck landing and catapult launch trials commenced with two Gannets, one of which had been modified with features that were later to be incorporated in the AEW version of this aircraft. The other aircraft was fitted with the Blue Silk Doppler Navigation System and Automatic Plotting Board, Mark 3. Landings and free take-offs were also carried out by the Seamew A/S Mk 1 aircraft.

At the beginning of December, 1955, Avengers of the Ship's Flight carried out a mirror deck landing trial. First results indicated that the Avenger could probably be cleared to use it with experienced pilots, but it was considered that further trials were required before full conclusions could be reached.

During the last week of the autumn period, an Avenger of the Ship's Flight achieved the 2,000th arrested landing since the start of flying in February, 1955. This did not include a count of an even greater number of 'touch and go's'.

The Christmas period was spent alongside at Portsmouth, during which a ship's company 'At Home' was held on board and 920 guests were entertained. Dancing to the ship's dance band took place on a specially laid dance floor. Also, whilst at Portsmouth, Air Commodore Wynter-Morgan, RAF (Retd) presented the ship with a silver model of a Seahawk on behalf of Sir W.G.

RMS **Queen Elizabeth** *berthed astern of* **Bulwark** *at Cherbourg.*

"Sea Enterprise"

820 Squadron's Gannets are launched.

845 Squadron's Whirlwinds return from a sortie.

A lucky escape for one of 820 Squadron's Gannets.

A visiting Dutch Avenger had problems on take-off. Luckily the crew were rescued.

Mr Wilmot, Secretary of the British Insurance Association, presents a plaque in commemoration of the ship's adoption by the B.I.A.

Armstrong Whitworth Ltd, makers, under contract, of Seahawk aircraft. The British Insurance Association made a further presentation in the form of a fine bronze plaque to be mounted on the quarter deck to commemorate the adoption of the ship by the Association.

Whilst at Devonport in February, 1956, the Commander-in-Chief, Plymouth, paid an informal visit to the ship.

Bulwark slipped from Devonport on Monday the 6th and when sufficient sea room had been gained, two large details of Seahawks were flown on carrier drill. During the final land on, the starboard oleo leg of a Seahawk of 897 Squadron collapsed on the deck. This was the second occasion on which this had occurred during the Squadron's embarked period. In view of a past history of similar failures, it appeared certain that the cause was structural weakness, so flying operations with aircraft of this mark—which constituted the whole of 897 Squadron—were suspended.

On Saturday 11 February *Bulwark* arrived at Bangor, Northern Ireland, anchoring in Bangor Bay. Due, however, to the lee shore, bad holding ground and the swell at Bangor Pier, making it too hazardous to land liberty men, the ship proceeded to Greenock. She remained there until Monday when it was necessary to return to Bangor to collect ground personnel of 812 Squadron from RN Air Station, Eglinton and three Skyraiders of 849A Flight from RN Air Station, Culdrose.

GANNET
over the side . . .

Another lucky escape. Sub. Lt. Allkins skids to port on take-off. . . and is rescued by ship's S.A.R.

The following weekend, whilst at Greenock, a Board of Inquiry was held into an accident which had occurred two days earlier. A Gannet of 812 Squadron was lost when, during take-off, the pilot was unable to keep straight on a wet deck after misalignment of the aircraft's nose wheel. The aircraft plunged over the port side after running two hundred feet up the deck. Fortunately the pilot (Sub-Lieutenant Allkins) was quickly recovered, unhurt, by the SAR helicopter.

On Friday 24 February the ship arrived at Belfast and proceeded alongside the Airport Wharf at Sydenham. This was *Bulwark*'s first return to Belfast since the start of the commission in November 1954. The usual parties took place and on the Sunday, the Captain presented a plaque to members of the Royal Naval Association, Bangor, in recognition of the hospitality which this excellent Association always extended to ships' companies visiting Bangor.

Bulwark left Belfast on Monday 27th and proceeded southwards—a ship's Avenger was flown to provide functional testing for the blind firing directors of the Bofors guns which had been 'uncocooned' for examination. The ship anchored overnight at Milford Haven and sailed for Portsmouth the following day. At 1030 the Dutch motor vessel *Rigel* was sighted, broken down and in distress, and, in response to her request for assistance, *Bulwark* took her in tow to Milford Haven.

The M.V. Rigel is taken in tow.

During the weekend 17-18 March a silver replica of a television camera was presented to the ship by representatives of the BBC, as a mark of appreciation for the help given by the ship in June 1955 when the first live television ever made from one of Her Majesty's ships at sea was produced and transmitted from *Bulwark*.

The ship proceeded to the Plymouth Exercise Area to meet the Home Fleet for exercise 'Dawn Breeze'. This exercise had no tactical setting but comprised a number of serials designed to exercise the passage of a convoy through submarine infested waters; fighter defence of a carrier task group; strikes against defended targets and a surface anti-raider action. On completion of the exercise, *Bulwark* was detached from the Home Fleet and returned to anchor in Sandown Bay.

Earlier in the year, scientists from the Admiralty Experimental Establishment at Haslar had re-installed a vertical motion recorder in the ship, to measure and record the motion of the flight deck. A previous trial had been carried out in *Bulwark* with this equipment during the summer months but there had been insufficient deck motion to provide all the necessary data. This time, in the Bay of Biscay, during exercise 'Dawn Breeze', all the necessary recordings were achieved!

751 Squadron in exercise "Dawn Breeze".

An unfortunate fact of life for a Trials and Training Carrier was that the ship was debarred from nearly all competitive sport with other units of the Fleet. Accordingly, as much accent as practicable was placed on inter-part sport. In 1955 the British Insurance Association had presented the ship with a very handsome silver trophy—expressly for inter-part competition on board. Aggregate points for the soccer, deck hockey, water polo, tug of war and ·22 rifle competitions were so close that the whole contest depended on the whaler race held at Portsmouth on 27 March. This was won by the junior Engine Room crew, which meant this group tied with the Chief Petty Officers and Petty Officers as joint winners. The trophy was presented by Mr Pollen, the president of the British Insurance Association who travelled from London especially for the event. In the evening, the BIA provided a magnificent supper for all the winning teams (62 ratings).

After Easter leave, *Bulwark* left Portsmouth on 28 May but embarkation of 890 and 893 Sea Venom Squadrons was delayed—by lack of wind. Flying was restricted to catapult launches and deck landings could only be achieved by Ship's Flight Avengers and by a Seahawk from Boscombe Down. On the 31st, shortly after the last of the aircraft had been embarked, a fatal accident occurred when one of the first detail of Sea Venoms to be launched from the catapult, crashed into the sea immediately on take-off. The planeguard helicopter was immediately over the wreckage and a ship's boat followed but no trace was found either of the pilot Lt E.L. Johnson, RN, or his observer, Lt G.M. Nichols, RN.

After a weekend in Torquay, *Bulwark* sailed to continue flying with 890 and 893 Sea Venom Squadrons. During the first detail though, a second fatal accident occurred when a Sea Venom flown by Lieutenant Commander P.S. Brewer, RN (the Commanding Officer of 890 Squadron) stalled just short of the round down, striking its wing on the safety nets below the batsman's position and, turning over, crashed into the sea. The body of Lieutenant Commander Brewer was recovered by the planeguard helicopter but searching helicopters, aided by the sea boat, found no trace of the observer Lieutenant E.E.J. Massee, RN. The ship proceeded to Plymouth where a Board of Inquiry was held into the accident.

On departure from Plymouth for Belfast, the wind was westerly and favourable for flying so the opportunity was taken to give carrier experience to Dartmouth Cadets in Sea Balliol aircraft flown by their instructors of the Dartmouth Flight.

It was not practicable to proceed alongside at Belfast on arrival because of high winds, so the ship first anchored off Bangor and berthed at Airport Wharf at Sydenham the next morning, Saturday 9 June.

Whilst at Belfast, His Excellency the Governor General of Northern Ireland made an informal visit to the ship.

Bulwark sailed from Belfast on 11 June and continued flying operations in the Irish Sea and off the Western Isles of Scotland with 815 and 849C Squadrons. A Gannet was lost on the 13th, when a premature break-out occurred on the port catapult, allowing the launching bridle to fall off and the aircraft to travel up the deck and plunge into the sea at about 30 knots. Fortunately the pilot, Sub Lieutenant D. Lewis Bowen was recovered unhurt.

After a visit to Buckie which ended on 24 June, *Bulwark* sailed and embarked aircraft of 895 Squadron before she left the Moray Firth and set course for Copenhagen. During a four day visit, tours of the Carlsberg and Tuborg breweries were arranged for—and much enjoyed by—the ship's company.

Between 23 June and 7 July three Squadrons of the Southern Air Division of the RNVR (Nos 1832, 1835 and 1836) carried out landings onboard. They were flying Attacker and Seahawk aircraft and became the first Reserve pilots to carry out landings on a carrier with jet aircraft.

On completion of a self-maintenance period alongside at Portsmouth, *Bulwark* proceeded on 16 July to Sandown Bay embarking Ship's Flight aircraft on the way. The next three days were taken up with the Commander-in-Chief, Portsmouth's inspection. To recuperate, the next weekend was spent at Bournemouth.

Bulwark anchored in Plymouth Sound on 25 July to embark 25 Admiralty representatives of a Committee set up to investigate the effect of the high aircraft noise level on internal communications in aircraft carriers.

During the ship's visit to the Isles of Scilly, the weather was most unfavourable, with heavy rain and high winds. In the early hours of Sunday 29 July orders were received to return to Portsmouth forthwith. In driving rain and a violent storm with windspeeds of 60 knots, great difficulty was encountered in extricating the ship from St Mary's Roads. During passage, several messages were received from, and about, ships and yachts in difficulties, but *Bulwark* requested permission to be released from searching because of the urgent need to disembark aircraft of 809 and 824 Squadrons before arrival at Portsmouth.

The sudden change in the ship's programme and the increased tempo of operations had been caused by the Egyptian Government, then under President Nasser, having decided to nationalise the Suez Canal. The main stock of the Canal had been in the hands of the British Government and French private investors. The British and French Governments hoped to force Egypt into accepting international control of the Canal, by the formation of a Suez Canal Users Association. This plan fell through when the Soviet Union vetoed the proposed Association in the UN Security Council. Meanwhile, the Royal Navy was preparing for all eventualities, and *Bulwark*'s programme was adjusted accordingly.

The remainder of that week was taken up with an intensive storing, ammunitioning, fuelling and embarkation programme in preparation for the ship's new operational role.

On 6 August, as the ship left Portsmouth, aircraft of 810, 804 and 895 Squadrons, a total of 30 Seahawks, together with the Ship's Flight, were embarked and she then set course for the Mediterranean.

Two days were spent in dry dock at Gibraltar, before sailing eastwards to Malta. An intensive work-up period was undertaken in company with *Eagle* and *Albion.* Strikes were flown against the two French carriers *Lafayette* and *Arromanches.* An enjoyable four days were spent at Marseilles at the end of September.

During flying exercises in the first week of October, Sub-Lieutenant Meredith of 804 Squadron was killed when he ejected from his aircraft because the fire warning light had come on. His body was recovered and on Friday the 5th, *Bulwark* was detached from exercises to bury him at sea.

The ship returned to Gibraltar on 13 October, where work was put in hand to rectify a serious defect in the starboard catapult—a crack in the casing of the draw-back valve. On the evening of Thursday 18 October orders were received to return to normal notice and sail with despatch for Malta as soon as the ship was fully operational. The Dockyard and ship's staff worked round the clock and the ship was able to sail at 1600 on Tuesday 23 October.

At 28 knots

Quite a bow wave . . .

. . . And a stern wash. ▶

Into the Mediterranean

Hands to bathe—at last the water's warm.

Entering Grand Harbour, Malta.

With **Eagle** *at Syracuse.*

Operations against Egypt

Armed and ready to go.

A couple of days were spent in Malta before *Bulwark* headed eastward at 25 knots, rejoining *Albion* and *Eagle* for operations against Egypt.

Such flying as was practicable, was carried out during Monday and Tuesday; a Combat Air Patrol was flown throughout Wednesday with aircrews standing by all day. Shortly after noon, ships' companies were informed that military action against Egypt would commence at 1500 local times and in the evening aircrews were briefed for pre-dawn strikes the following day.

On 31 October 1956 after Egypt's rejection of an ultimatum by the British and French Governments, joint Anglo/French amphibious operations began, synchronised with a land offensive by the Israelis.

In the early stages of these operations (codenamed 'Musketeer') a further fatal accident occurred when a Seahawk crashed while landing on *Bulwark* because the arrester hook pulled out. Unfortunately the pilot was lost.

During the whole operations up to the ceasefire, which occurred at 2359 on 6 November, nearly six hundred sorties were flown by *Bulwark*'s aircraft. Her Seahawks had supported the troops ashore; destroyed a number of tanks; knocked out a gun battery at Port Said and sunk three Motor Torpedo Boats. During the initial stages of the campaign, the Egyptian Air Force had been virtually destroyed on the ground by the carrier-borne aircraft.

Due to serious defects in both *Eagle*'s catapults the Flag Officer Aircraft Carriers transferred his Flag to *Bulwark* at 1700 on Wednesday 7 November and remained in *Bulwark* with his operational staff until Thursday 15 November, when he returned to *Eagle*. *Bulwark* sailed from the operational area pm on Friday 16 November for a brief self maintenance period in Malta, mainly necessitated by the need to make good a serious defect in the port catapult extension cylinder joints.

Before leaving Malta on Saturday 24 November to return to the operational area, two Wyverns and three Sea Venoms were embarked as replacement aircraft for *Eagle*, bringing the ship's total, including helicopters, up to forty-two aircraft, the largest number carried in her career to date.

Eagle's replacement aircraft were flown to her on Monday 26 November and *Bulwark* returned to the operational area early on Tuesday 27 November. However by this time measures were being taken to reduce the intensity of the military effort. The Governments involved had bowed in the face of worldwide displeasure and were to withdraw their forces without achieving their objectives. The dispute ended with the signing of an agreement under which all British forces were to be withdrawn within 20 months and the canal was to be under Anglo-Egyptian control.

The ill-fated Suez Campaign had confirmed the role of the aircraft carrier in a 'limited war', the British carrierrs having maintained an excellent rate of sorties—with very few accidents.

Seahawk over Nile Delta.

The Governor's Palace burning after attack by ship's aircraft. ▶

Attack on Helwain airfield.

The Homecoming

After the war . . . comes "Divisions". Malta, December, 1956.

In company with *Eagle* the ship returned to Malta on Saturday 1 December. While on passage a routine examination of the starboard catapult rope revealed that eighteen strands had parted and a mandatory change of rope was necessary. Work, which was carried out in shifts, was completed on Wednesday 5 December.

On Sunday 9 December the Flag Officer Aircraft Carriers inspected divisions on the Flight Deck and afterwards addressed the ship's company; attended Divine Service on the Quarter Deck and later talked to officers in the Wardroom. In the afternoon the ground party of 812 Squadron was embarked and the baggage and Squadron mobile equipment of 804 Squadron disembarked to Hal Far to await the Squadron's return to Malta.

Bulwark sailed from Grand Harbour at 0900 on Monday 10 December to the accompaniment of bands and cheering parties from ships in Grand Harbour and with many kind and complimentary messages of farewell. 804 Squadron aircraft were launched for Hal Far and the pilots were subsequently re-embarked by helicopter.

Twenty-four hours were spent at Gibraltar to allow the remaining stocks of Avgas to be discharged, so that the tanks could be washed through while on passage to Spithead. The ship sailed from Gibraltar on Friday 14 December. The passage to Spithead was uneventful except for some seasonal rough weather.

Off the Lizard on the 17th, aircraft of 810 and 895 Squadrons were flown off to Lee-on-Solent. 'Finished with main engines' was ordered for the last time in the commission on 18 December, bringing to an end an eventful first commission which had started in the builder's yard and ended after the active service of the Suez Campaign.

3

Second Commission 1957 to 1958

On 15 January 1957 Captain Villiers handed over command of *Bulwark* to Captain P.D. Gick, OBE, DSC* Royal Navy. Whilst still in dockyard hands at Portsmouth, *Bulwark* recommissioned on Wednesday 20 March 1957. Easter leave came and went but it was not until the ship was clear of dockyard workers that the official Re-Commissioning Service was held. The Chaplain of the Fleet officiated on the flight deck during the afternoon of 12 June. After the service there was tea and dancing in the hangar.

Her departure, for a preliminary 'shake down' before embarking her Squadrons, scheduled for two days later, was delayed for 24 hours owing to a blown joint in the superheated steam line.

The following weeks were taken up with full power and steering trials; and radar and gunnery calibrations. A surprise visit was made to Bognor Regis and a week-end was spent in Portsmouth before putting to sea to embark on 25 June, 891 (Sea Venom) Squadron; 820 (Gannet) Squadron; and 849 (Skyraider) Squadron, 'D' Flight.

Three days later *Bulwark* rendezvoused at sea with *Ark Royal*—who was returning from a visit to the USA—and, from her, embarked 898 (Seahawk) Squadron. Stores and personnel were transferred by helicopter in less than an hour.

The last weekend in June was spent at anchor off Milford Haven. Leaving there to begin night flying, a large number of fuel gauges in the Gannets and jets were discovered to be unserviceable due to water contamination of the Avcat system. All Gannet and jet flying was cancelled to allow the aircraft tanks to be drained and the ship's Avcat system flushed. Limited flying was resumed on completion, but the symptoms appeared again. The immediate programme and the exercise planned for the following week were cancelled and the ship anchored in Bangor Bay, Northern Ireland, to carry out the lengthy task of purging the entire Avcat system.

In Bangor Bay a minor sensation was caused when, with the ship still at anchor, a Gannet was catapulted. It broke from the catapult and trundled into the sea, from where its three occupants were rescued by excited holidaymakers who had been clustering round the ship in an assortment of small boats. To the uninformed onlookers, the Navy's behaviour must have seemed eccentric in the extreme: *Bulwark* arrived at Bangor, anchored, 'ditched' a Gannet into the sea, weighed her anchor and sailed off to Belfast.

After a period of self-maintenance at Belfast, *Bulwark* sailed on 15 July for Exercise 'Bullcan', the postponed anti-submarine work-up. After its completion she, in company with the frigate *Salisbury*, arrived at Cherbourg on 25 July for the first foreign visit of the new commission.

Five days later she sailed from Cherbourg and, in spite of adverse weather conditions, was able to complete some deck-landing trials for the French 'Fouga' trainer aircraft. Then, after an unexpected week-end at St Peter's Port, Guernsey, the ship arrived back in Portsmouth on 6 August for summer leave.

After embarking 845 Helicopter Squadron on 28 August, the ship sailed north to join the rest of the Carrier Squadron at Rosyth. In company with *Eagle* and *Ark Royal*, intensive day and night flying was carried out until arrival in the Clyde on 13 September. There the carriers joined the remainder of the NATO Forces assembled there for Exercise 'Strikeback'.

Between the 17th and 28th of the month, over 35,000 men (12,000 of them British), 100 ships and 700 aircraft operated in a large area of the north-east Atlantic, stretching from within the Arctic Circle to points near the Equator.

The weather was clear and sunny but *Bulwark*'s ship's company was hit by the ravages of Asian Flu which took so great a toll of victims that messes had to be converted into emergency isolation wards. All the ventilation was turned on to full power and PT was organised each day on the flight deck in efforts to stave off what came to be known as the 'dreaded lurgy'.

As the exercise drew to its close the ship returned to Belfast much to the indignation of the 'Pompey' natives. A ship's dance was held and there was a full sports programme before *Bulwark* sailed on 11 October. An unscheduled visit was made to Penzance whilst en route to the rendezvous with ships taking part in Exercise 'Rum Tub'. Towards the end of this exercise, the ship achieved her largest roll to date—27 degrees!

The most popular foreign visit of this part of the commission followed when, having disembarked 845 Squadron

Docked again at Gibraltar.

Helicopters "escort the Escort".

Suez revisited—the war now forgotten and the canal clear.

S.S. **Talodi** *left her mark during the night.*

Refuelling and yet more refuelling.

off Spithead, *Bulwark* headed for Rotterdam, arriving there on 22 October.

A long awaited self-maintenance period took place at Portsmouth between 6 and 16 November. The ten days soon passed, 898 Squadron was re-embarked, along with 801 Squadron Sea Hawks. These together formed the main part of the ship's air group during Exercise 'Phoenix II'.

After Christmas leave and a hectic storing programme, *Bulwark* left Portsmouth on 10 January. Five days later she joined the Home Fleet at Portland for that Fleet's cruise to the Caribbean. In addition to exercises with the USN and RCN, there were enjoyable visits to various ports in the area. *Bulwark* became the first RN ship ever to visit Cannouan Island in the Windward Isles when she anchored there to have four days respite from flying. To a people who had only ever seen cars in magazines, the arrival of *Bulwark* with her aircraft and helicopters caused quite a sensation.

After visits to Jamaica and Bermuda, *Bulwark* rejoined the rest of the Home Fleet and the RCN Atlantic Fleet. The whole force formed up into two columns for its entry into the historic port of Halifax.

Halifax was covered with a blanket of melting snow, but this cool reception was more than offset by the hospitality of RCNAS *Shearwater*, who threw open their establishment to *Bulwark* for the whole of her stay. The ship sailed for Exercise 'Mayroy II' on 18 March.

However, three days later she was detached to Gibraltar, where she arrived on the 26th to enter No. 1 Dock for a 10 day self-maintenance period.

The afternoon of Easter Monday saw *Bulwark* set out on the long passage to the Far East. On 13 April, whilst passing through the Suez Canal, the tips of a blade on each propeller were slightly bent on an unidentified obstruction. The ship cleared the Canal at 1930 and made for Aden. At 0330 the following day the ship's company were woken by a scraping noise—the ship had collided with an Egyptian ship SS *Talodi.* The merchantman had struck *Bulwark* a glancing blow on the starboard quarter, piercing the hull in several places. *Talodi*'s bows were flattened for about 15 feet, she lost her anchors and her foremast, and the fact that she was carrying about 120 pilgrims to Mecca added to the confusion. After ensuring that the *Talodi* was still seaworthy, *Bulwork* went on her way, her own damage being superficial and well above the waterline. It was made good after her arrival at Aden on the 16th.

On 28 April, 41 days after leaving Halifax, *Bulwark* berthed alongside HM Naval Base at Singapore, where a combined SEATO Force was assembling. Three days later, wearing the Flag of Rear Admiral Durlacher, FO2FES, she sailed for Exercise 'Ocean Link'—the first phase of the largest SEATO exercise yet held, which would involve ships of five nations. In overall command was Rear Admiral Almon E. Loomis, of the US Navy, flying his Flag in USS *Philippine Sea.* The object of the exercise was to improve the efficiency of combined naval operations in the SEATO area and to promote closer working relationships between the various nations to improve mutual defence.

Led by *Bulwark,* the SEATO fleet entered Manila Bay during the forenoon of 13 May for a short visit. Four days later she berthed at the north jetty of HM Dockyard, Hong Kong.

Bad weather delayed the ship's departure from Hong Kong for Exercise 'Sea Horse' for 24 hours. She finally sailed on 3 June. Apart from the period at Portsmouth over Christmas, 1957, this had been *Bulwark*'s longest stay in any port since August 1957.

The Far East Fleet—of which *Bulwark* was now a unit—joined the East Indies, Indian and Pakistani Fleets for the first phase of a Joint Exercise 'Foamite'. On completion, on 20 June, the whole force arrived at Trincomalee. Owing to a state of emergency in Ceylon at the time, a curfew was in force and leave was restricted to the dockyard area. After a quiet weekend the combined fleets sailed for the second phase of the exercise, which was appropriately named 'Extinguish'.

With the frigate *Ulysses, Bulwark* parted company from the combined fleets at Trincomalee and made for Mombasa. On passage, on 4 July, she crossed the equator for the first time in her career.

That evening the ship was visited by a herald to King Neptune, announcing the ceremony that was to begin with the arrival of King Neptune and his Court at 0900 the next day. A large number of the ship's company were initiated after nearly all the senior officers had been "reprimanded and punished for heinous crimes".

Whilst at Mombasa, a group of seven stalwart members of the ship's company decided to attempt to climb Mount Kilimanjaro, at 19,340 feet the highest peak in the African continent, and about 100 miles inland from Mombasa. After four days' sweat and toil, of blistered feet and sunburned faces, four out of the seven reached the summit. Towards the end of this visit, a crisis arose in the Middle East and, instead of sailing for Madagascar as planned, she embarked troops and equipment and sailed for Aden on 20 July.

What every well dressed sailor should wear on Kilimanjaro.

Preparations for self-maintenance at Aden were abandoned and August Bank Holiday Monday found the ship, again in company with *Ulysses*, conveying the 1st Battalion Cameronians and their equipment to Aqaba where they were to safeguard the British supply line from Aqaba to Amman.

The overdue self-maintenance eventually took place at Aden from 10-19 August. When the ship put to sea for further flying exercises, the very first launch resulted in a mishap—a Seahawk ditched on being catapulted. The pilot was recovered suffering from shock and minor injuries, and was flown to the RAF Hospital at Steamer Point.

On Sunday 31 August, a Joint Parade was held on board to mark *Bulwark*'s friendship with the 1st Bn the Royal Lincolnshire Regiment. The salute was taken by H.E. the Acting Governor of Aden, K.W. Simmons, CMG.

The following month *Bulwark* carried out flying operations against the rebel forces holding the 10,000 feet mountain range to the west of Muscat in the Jebel Akhdar. These were interrupted when at 0700 on 13 September messages were received that the French tanker *Fernand Gilabert* and the Liberian tanker *Melika* had collided earlier that morning. An account of the transfer of survivors to hospital and the subsequent salvage of both ships is recounted below.

Bulwark returned to Aden on 24 September and after a hectic but jubilant three day period of embarking stores and fresh provisions, sailed for the Mediterranean. A daylight transit of the Suez Canal was made on 30 September 1958. In mid October she was involved (with HM Ships *Eagle, Albion* and *Sheffield*) in emergency operations off the Lebanon and Jordan but was released in time to arrive back at Portsmouth on 4 November. This was to prove to be her last commission as a fixed wing aircraft carrier.

A remarkable salvage operation

"One of the finest salvage operations of recent times" was how the Admiralty described the work of the Royal Navy in saving two tankers which collided and caught fire in the Gulf of Oman. *Bulwark*'s role must surely rank as one of the most memorable incidents in the ship's life.

The Liberian tanker *Melika* 20,551 tons, and the French tanker *Fernand Gilabert,* 10,715 tons, collided on the night of 12/13 September, 1958 and both caught fire. *Melika* was carrying 23,000 tons of oil, and when the flames swept her decks, burning oil spread over the sea.

At 0700 on Saturday the 13th, distress signals were picked up by *Bulwark* who was exercising about 150 miles away with the frigate *Loch Killisport.* They made for the scene at 23 knots, along with the frigates *Puma* and *St Bride's Bay* who were also in the area.

When one of *Bulwark*'s Skyraiders arrived she found the *Fernand Gilabert* enveloped in smoke and flame, and down by the bow. The Swedish tankers *Ceres* and *Sira* and the British tankers *Anglican Diligence* and *Border Hunter* were standing by the stricken ship. They were picking up survivors. The injured among them were later transferred to *Bulwark* by helicopters of 845 Squadron. The Executive Officer, Commander R.H.H. Brunner, and a small fire-fighting team were transferred to the French ship.

Helicopters are used to ferry salvage equipment.

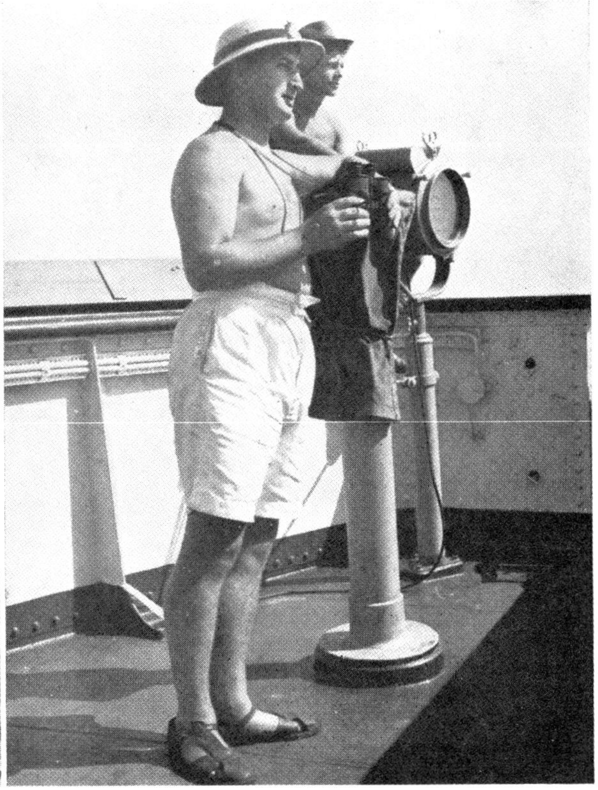

Lt. Cdr. Dane as temporary master of **Melika.**

The *Melika,* although abandoned, continued to steam southward at speed after the collision. When she was spotted, she was on fire amidships and listing. *Bulwark* closed with her and put a party of four on board, in charge of the First Lieutenant, Lt Cdr P.P.R. Dane.

Bulwark then steamed towards Masira to fly the injured to the RAF base there for onward flight to Bahrein and hospital. When she got back to *Melika* she found that the *Puma* had put a fire party on board the after end of the tanker but they were separated from *Bulwark*'s fire-fighters by the flames amidships. Meanwhile the party on board *Fernand Gilabert* radioed that all fires were out. The *St Brides Bay* took her in tow but eventually had to abandon the tow because of the poor weather conditions. Ultimately *Loch Killisport* towed her into Karachi.

The *Melika* after many trials and tribulations was towed into Muscat harbour by *Bulwark.* An Admiralty communiqué graphically describes events:-

> "Ablaze amidships, her gunwhale awash and seas forming geysers through the mid-deck hatches, *Melika* made an impressive sight as *Bulwark* prepared to pass a tow. In the moderate swell, both ships were rolling considerably, making boat work difficult. On one occasion a seaboat was caught under *Bulwark*'s counter, injuring two of the boat's crew, and on another *Melika*'s bows holed *Bulwark* above the water line. The tow was at last secured but progress was slow as *Melika* yawed widely. After only two hours the tow parted."

However the tow was again passed and *Bulwark* towed

*The other casualty—***Fernand Gilabert.**

her, while *Puma* secured a line to the tanker's stern and acted as a rudder. At Muscat *Bulwark* was relieved by the cruiser *Sheffield.* There, oil from the *Melika* was transferred to other ships, and ratings from the frigate *Loch Alvie* spent six weeks clearing up the ship. Temporary repairs were carried out by welding large steel beams into position and about 80 square feet of plating were used to cover the hole in her side to prepare her for the next stage of the tow to Palermo in Sicily.

The Naval tug *Warden* took the tanker in tow for the long, slow haul to the central Mediterranean. Escorted by the salvage ship *Sea Salvor,* the 3,800 miles voyage ended when the *Melika* reached Palermo early in December.

The fine feat of seamanship, ship-handling and ingenuity displayed in the actual evolution of getting the *Melika* in tow formed an interesting programme shown on BBC Television. It was filmed by photographers in *Bulwark.*

For their part in this salvage operation, 845 Squadron were awarded the Boyd Trophy for 1958. It was accepted by Lt Cdr H.M.A. Hayes and Lt Cdr C.M.A. Wheatley. The former had been in command of the Squadron till just before the incident, and the latter had been temporary Commanding Officer of the Squadron at the time.

. . . and so to a safe anchorage.

Conversion to Commando Carrier Jan 1959-Jan 1960

The landing of 45 Commando at Port Said in 1956 by an improvised helicopter force drawn from the three Services and operating from training carriers, reinforced the demand for a proper British helicopter assault carrier with dedicated helicopters and troops.

To meet this requirement it was decided to convert *Albion* and *Bulwark. Albion* was due to complete another 'Fixed Wing' commission, so *Bulwark* was the first to be converted.

In January 1959 work commenced in Portsmouth Dockyard to remove her fixed wing capability, including the arrester wires and catapults. During the next twelve months, various alterations were made to make the ship into an all-helicopter troop carrier carrying sixteen Westland Whirlwind helicopters. The alterations were such that the Whirlwinds could be replaced at a later date by Wessex helicopters. Eight 40mm A.A. guns were removed to provide space for four L.C.A.s carried at built-in gantries. The Operations Room facilities were modified to provide command arrangements for the military force. Extensive air-conditioning was fitted to enable the ship, with a 600 strong Commando on board, to operate in the most exacting tropical conditions. To install the steam turbine air-conditioning and refrigeration plant, several plates had to be removed from the ship's port side.

Bulwark became officially a 'Landing Platform—Helicopters', 'LPH' for short, but was generally referred to as a Commando Carrier. She could, however, be fully converted to fulfil an anti-submarine role, being able at short notice and within her own resources, to adapt her helicopters for A/S work.

When completed she was the first of her kind in the Royal Navy, and reinforced the traditionally close association of the Corps of the Royal Marines with the Royal Navy. She was to give these versatile troops greater mobility and support them during operations and exercises.

In conditions of limited war *Bulwark* would be able to provide a highly mobile amphibious force—a Commando of 600—and also be capable of embarking an additional Commando or Army unit of equivalent size for short periods, as a striking force to secure a beach and its immediate hinterland.

No. CO 2/60
MEMORANDUM

By Sir Manley Laurence Power, Knight Commander of the Most Honourable Order of the Bath, Commander of the Most Excellent Order of the British Empire, Companion of the Distinguished Service Order, Admiral in Her Majesty's Fleet and Commander-in-Chief of Her Majesty's Ships and Vessels employed and to be employed within the Portsmouth Naval Command.

H.M.S Bulwark — Commissioning Order

The Lords Commissioners of the Admiralty having directed that Her Majesty's Ship Bulwark is to commission at Portsmouth on Tuesday, the nineteenth day of January, 1960, you are to proceed forthwith to commission her and prepare her for a Foreign Service commission on the Far East Station.

On commissioning Her Majesty's Ship Bulwark will remain under my administration until further orders and you are to bring to my immediate notice any reason you may have for dissatisfaction with the general state of the ship or with any part of her. Similarly you are to bring to my notice any other matters of importance, in particular those relating to the welfare of your Ship's Company, so long as the ship is in the Portsmouth Command.

Given under my hand, this 18th **day of** January **, 1960**

L. Power

Admiral
Commander-in-Chief

To : Captain R. D. Franks, D.S.O., O.B.E., D.S.C., Royal Navy
H.M.S. Bulwark

4

Third Commission

1960 to 1961

The Commando Carrier *Bulwark* commissioned on 19 January 1960 under the command of Captain R.D. Franks, DSO, OBE, DSC, RN, who had actually been appointed to the ship in the previous August. The religious ceremony—which was held in the hangar because of the inclement weather—was conducted by the Venerable Michael D.S. Peck, MA, Archdeacon of Portsmouth, at 1100 on Saturday the 23rd.

Captain Franks' orders were to prepare the ship for a foreign service commission on the Far East station. Her work-up and trials were carried out in the Portland area, culminating in a four day Exercise 'Squarebash' with ships of the 7th Destroyer Squadron, RFA *Tidereach* and helicopters of 737 Squadron.

Monday, 22 February found the ship heading for Portsmouth and a week's foreign service leave. During the leave period the advance parties of 42 Commando began arriving on board, bringing with them much of their equipment and transport. Several important visitors inspected the ship at this time, including the Minister of Defence, the First Sea Lord, the Fourth Sea Lord, the Commander-in-Chief, Portsmouth, and the Admiral Superintendent, Portsmouth Dockyard.

On 10 March, shortly after leaving Portsmouth, the 16 Whirlwind helicopters of 848 Squadron, commanded by Lt Cdr B.M. Tobey, arrived on board. 42 Commando RM under Lt Col R.D. Crombie, RM joined the ship at Devonport with fitting ceremony. The Lord Mayor of Plymouth, Councillor P.N. Washbourn, with Lieutenant General I.H. Riches, Commandant General Royal Marines, took the salute as 42 Commando exercised the privilege of the Royal Marines as Honorary Freemen, marching through the city with colours flying, drums beating and bayonets fixed. At the dockside, Lord Carrington, First Lord of the Admiralty, watched the ceremonial embarkation, before the ship sailed for the Mediterranean.

After a three day stay in Gibraltar and a weekend in Malta, the ship started operational trials on the North African coast in the area of the small town of Homs. For the next few days the helicopters of 848 Squadron were busy landing the Commando and stores, and flying hours and numbers of landings soon reached an impressive total.

Over the weekend of 2 and 3 April, *Bulwark* was honoured by a visit from the Chief of the Defence Staff,

The "new look" **Bulwark** *at Gibraltar, her flight deck crowded with Commando transport.*

Admiral of the Fleet Earl Mountbatten of Burma. He arrived by helicopter and left the next day having inspected the ship and its assault organisation. The G.O.C., Malta, was another welcome visitor at this time.

On Monday 25 April, the ship returned to Marsaxlokk to be honoured by a visit from H.R.H. The Duke of Edinburgh. He met the Chief and Petty Officers and Senior NCOs, and later dined in the Wardroom. The following day he watched a rehearsal of 'Sky Pioneer' and later inspected the ship before transferring to the frigate *Eastbourne*.

By arrangement with the Libyan Government, Exercise 'Sky Pioneer', the final exercise in the Mediterranean

Work up off the coast of Libya

Briefing of Pilots and Commandos before the exercise.

A "stick" of Royal Marines running to embark.

Airborne . . . and heading for the coast.

The "Stick" leader raises his arm to inform the pilot that all his men are out.

Joint operations. Sailors and RM's washing sand from the island. The latter seem to be going faster—but better??

work-up of *Bulwark* and 42 Commando, took place on the coast of Libya, east of Tripoli, on 29 and 30 April.

After six weeks in the Mediterranean *Bulwark* sailed on 12 May for the Far East. After calling at Aden and Colombo, she arrived in Singapore on 7 June, to be welcomed by the pipers of the 99th Gurkha Brigade. The Commandos made their arrival known by staging an assault on the barracks at Sembawang. This former Royal Naval Air Station had been used as a Naval aircraft repair yard during the Korean War but in recent years had been occupied mainly by the Army. With the ship due to spend four weeks in dry dock, the ship's company moved ashore into the local shore base, HMS *Terror*.

848 Squadron and 42 Commando gave an impressive demonstration of their various capabilities to an audience of Senior Officers and other guests on Tuesday 12 July in Exercise 'Skylight'. That afternoon the ship invited about 300 children to a party on board, and they had a grand time on the flight deck riding in a train, an aeroplane and other equipment, followed by tea and a film.

On 15 July, with the ship's company and Commando re-embarked, the ship sailed for the Persian Gulf—for hot weather trials—but world events caused a change of plans, and the ship headed for Aden instead.

On the way, she called at Male, the capital of the Maldive Islands, firing and receiving 21 gun salutes in honour of His Highness the Sultan of the Maldives, whom *Bulwark* had the privilege of conveying to Colombo.

At Aden, vehicles and personnel were loaded and the ship was soon off to Mombasa.

At 1139 local time on Monday 1 August, *Bulwark* 'crossed the line' on her journey south.

During two weeks spent at Mombasa the ship was open to visitors and the local press reported that over 5,000 of the population took the opportunity to have a look over the ship. 42 Commando exercised ashore but rejoined

At Aden—en route to the Far East.

At Colombo—amongst the crowded shipping in the port.

before the ship sailed on 15 August for night and day exercises, and assault training off Malindi.

During these exercises, the ship's motor cutter was sunk but luckily the crew escaped with nothing worse than a wetting. The ship's company were engaged in general drills—what sights were seen and what ingenuity was shown! Sultans' harems appeared on the quarter deck, showers on the flight deck, and the Chinese cooks and stewards manufactured an elephant.

Reluctantly, the ship left Mombasa on Monday 29 August after a very pleasant stay. That evening King Neptune's herald arrived on the flight deck to announce that a full 'crossing the line' ceremony would be held on the following morning. The great day was bright and cheerful to greet the arrival of Neptune, his Queen, court officials and bodyguard. A procession of miscreants were brought before them, led by the Captain, the Colonel, and other Senior Officers. Regardless of plea or defence these were all 'treated by the surgeon, shaved by the barber and thrown to the bears'.

From 15-28 September *Bulwark* was alongside at Singapore, allowing the ship's company to relax with the usual shore-going and sporting activities.

At 0830 on Thursday the 29th the ship sailed for a two-day anti-submarine exercise. For this she carried the Wessex helicopters and crews of 815 Squadron from *Albion* and in company with the Australian destroyer *Vampire,* Australian frigate *Quiberon,* New Zealand frigate *Rotoiti* and HM submarines *Tactician* and *Teredo.* On completion *Bulwark* returned, not to the Naval Base, but to Singapore itself, anchoring in the Roads, where she remained for the weekend.

On the Monday morning the 2nd/2nd Gurkha Regiment were embarked and conveyed to North Borneo for training. There on 7 October the ship's landing party, known onboard as '*Bulwark*'s Light Horse', were flown ashore, their task being to defend a hill against an attack by the Gurkhas. After a gallant stand against overwhelming odds, the 'BLH' were forced to surrender. After being taken prisoner they were marched five miles, in a tropical

First arrival at Singapore.

Crossing the line . . . another victim is thrown to the bears.

downpour, to a temporary camp before re-embarking into *Bulwark* the next day.

On passage to Manila, *Bulwark* just missed Typhoon 'Kit', which caused much damage on shore and loss of life. However, she caught the fringe of Typhoon 'Lola', that luckily passed just north of the ship. Boat traffic was suspended as she rode out the gale at anchor in Manila Bay, with visibility cut right down by heavy rain. Two days of the visit were spoiled because of this weather.

Monday 24 October was a proud day for the ship when the Flag of F.O.2.F.E.S., Rear Admiral M Le Fanu, CB, DSC, was transferred to her. He was to remain on board throughout the forthcoming visit to Hong Kong.

Early in November whilst on passage, with 42 Commando on board again, *Bulwark* conducted various exercises with other HM ships and RFAs. Once the ship arrived in Hong Kong, 42 Commando were quickly landed to exercise in the New Territories with the resident Army unit.

December was spent at Singapore for a self-maintenance period. On completion, the ship headed for Mombasa for a two week visit. There the GOC, East Africa, Major-General Goodwin, CB, CBE, DSO, paid an official visit to the ship and separate groups from the ship's company and Commando successfully scaled Mount Kilimanjaro.

Visits to Bahrein and Muscat followed before she returned to Aden on 24 February. *Bulwark* left there for Exercise 'Roulade' on Friday 3 March and steamed about 100 miles up the coast to begin an assault at 0500 the next day. This was carried out by the ship's Commando aided by ships of the Amphibious Warfare Squadron.

On Wednesday the 8th, the ship was honoured by a visit from the First Sea Lord, Admiral Sir Caspar John, GCB.

After a 'short weekend' at Aden, the ship returned to Singapore, where frenzied preparations were made for the Admiral's Inspection, carried out by F.O.2.F.E.S. Rear Admiral Le Fanu, CB, DSC, on 25 March 1961.

A few days' anti-submarine exercises were carried out before preparations were put in hand for a major exercise, code-named 'Pony Express', on the coast of North Borneo. Three months of intensive planning lay behind this, the biggest military exercise ever planned in the 6½ years of the South East Asia Treaty Organisation (SEATO). It involved about sixty ships from six nations carrying out a coastal bombardment, followed by an assault from the sea

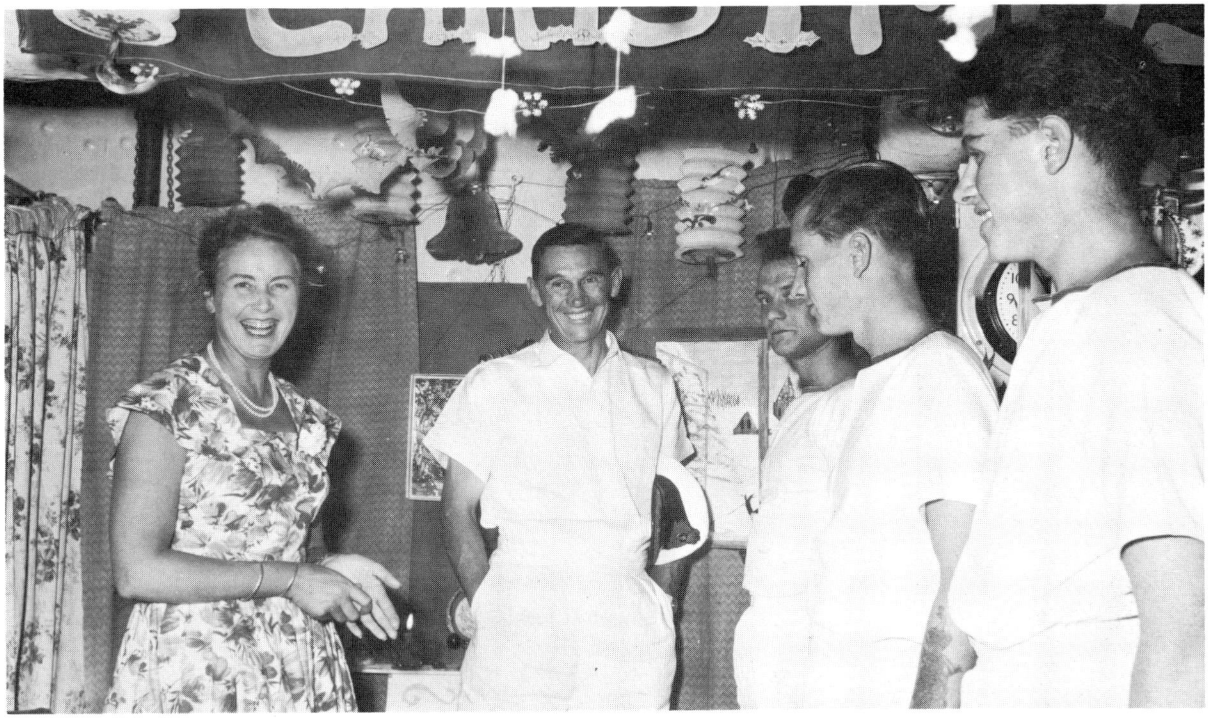

Christmas in Singapore. Captain and Mrs Franks visit the mess-decks.

At sea again—the Commando keep fit.

Celebrating the 5,000th deck landing during the commission.

and by air. Six thousand men were landed under battle conditions on the North Borneo coast. The British ships *Victorious*, *Bulwark*, *Belfast* and two Squadrons of minesweepers and support craft, took part along with the New Zealand cruiser *Royalist* and US Assault Helicopter Carrier *Thetis Bay*. 42 Commando, RM, Australian Infantry and US Marines formed the assault force. Attacks on shipping were made by British and US submarines, while RAF bombers from Singapore acted as enemy aircraft. The exercise ended on 4 May, and after a few days in Singapore, *Bulwark* sailed on the 10th for a 'Showing the Flag' cruise. Visits were made to Sattahib in Thailand; Cape St Jacques, 30 miles from Saigon in South Vietnam; and Labuan Island off the north west coast of Borneo. The ship was visited by the Sultan of Brunei and gave demonstrations of the ship's capabilities at Kuching and Sibu (Sarawak) and Tawau (British North Borneo).

Shortly after leaving Tawau *Bulwark* carried out anti-pirate operations but the result could be put down to a victory for the pirates for none were sighted!

Eight days were spent in Hong Kong before returning to Singapore where the ship was due to recommission. The advance party for the new commission arrived on board five days later and the following day those men selected to go home early left Singapore Airport.

Crisis in Kuwait

However, on 21 June *Bulwark* sailed for hot weather trials in the Persian Gulf, but, just after refuelling at Karachi, news was received of an impending crisis in the tiny Sheikhdom of Kuwait.

The Iraqi Prime Minister, General Kassim, had revived an old Iraqi claim to Kuwait, having stated at a stormy press conference that Kuwait was, and always had been, legally an integral part of Iraq. He rejected the Anglo-Kuwait Agreement of 1899 and also the more recent agreement contained in an exchange of notes between the UK and Kuwait. Rumours were rife that Iraqi troops were being moved southward from Baghdad to the Kuwait border, so in the face of Iraqi threats, the Sheikh mobilised his own tiny army and declared a State of Emergency.

On 29 June Great Britain alerted air and ground forces for movement to the threatened Sheikhdom. *Victorious* was summoned from the Far East and *Bulwark* appeared

off Kuwait within 24 hours of the formal request for assistance being made. Immediately the 600 men of *Bulwark*'s 42 Commando were transferred ashore by helicopters. By nightfall they had secured an uncompleted airstrip ten miles outside the city of Kuwait so that it could be used as an air base.

The build-up of British forces in Kuwait continued by sea and air from bases at Bahrein, Aden and in Kenya, and from the strategic reserve in UK. The objective, though, was to provide a deterrent and thus avoid combat.

By 8 July there were signs that the intervention had succeeded. Iraq's Prime Minister refused to withdraw his claim to Kuwait but gave assurances he would not try to enforce it by arms.

The "Helicopter Control Team" ashore.

Operating off Kuwait—the initial assault.

Return to Singapore . . .

After three weeks off Kuwait, *Bulwark* re-embarked her Commando and sailed for Singapore. With her paying-off pennant flying, helicopters overhead and 'escorted' by her own four landing craft, she arrived there on 31 July, bringing to an end her first Commando commission during which she had steamed over 64,000 miles but, more importantly, had proved and established herself in her new role.

Visitors. Famous faces seen onboard during this commission

Lord Mountbatten meets former shipmates.

His Royal Highness signs a photograph to hang in the mess.

Mr Edward Heath comes on board, too.

5

Fourth Commission
1961 to 1962

Paya Lebar airport—the commission started here.

In 1961 *Bulwark* became the first RN aircraft carrier since World War II to recommission away from her UK Base Port. This entailed about 900 officers and men being flown from UK to Singapore. On arrival, they were accommodated in HMS *Terror* commuting daily to the ship whilst she completed a short refit.

On Saturday 7 October 1961 the ship was formally commissioned under the command of Captain J.S. Dalglish,

With the ship in dock . . .

. . . HMS **Terror** *provides hotel services.*

Orphans' party at Hong Kong.

CVO, RN, who had been appointed to *Bulwark* on 6 August. The Commissioning Warrant was read by Commander H.G. Vere, DSC, RN.

After completion of post-refit trials, a weekend was spent at the East Malaysian island of Pulau Tioman where the ship's side was painted. When the ship arrived in Hong Kong in early December, that task was performed by the famous 'Jenny's Side Party' (a group of about 20 Chinese ladies operating from sampans who paint the ship using long-handled brushes). Jenny and her ladies have been carrying out this task for visiting RN warships for many years. Later her work was recognised by the award of the British Empire Medal.

Christmas was spent in Singapore and at 1600 on 4 January 1962 *Bulwark* sailed south and made a daylight passage through the Gaspar and Sunda Straits. On the 7th, for the first time during this commission, she crossed the equator with all the usual ceremonies.

Five days later the ship arrived off the small town of Geraldtown, about 300 miles north of Perth in Australia. 3 Commando Brigade, RM and their vehicles were to land for an exercise (codenamed 'Tombstone') with the Australian Army. A heavy swell prevented the use of LCAs so they were disembarked by Whirlwind helicopters instead.

On 12 January the ship arrived at Fremantle, where on the 16th, 300 of 42 Commando, with their light equipment, and about 500 of the ship's company, were privileged to march through the streets of Fremantle and Perth. The march was led by the ship's Volunteer Band, backed up by the Australian Western Command Band. The Commando paraded their Regimental Colours with a guard, and the rest of the Commando were in battle order, including support troops in vehicles. In Perth, the Governor, Sir Charles Gairdner, took the salute outside Trinity Church. Also on the dais was the Lord Mayor, Sir Harry Howard. Overhead, helicopters of 848 Squadron flew past in formation.

Every opportunity was taken by the Australian military forces to make the most of the presence of *Bulwark*. Exercises ranged from the helicopter training of Australian Special Air Service troops, including parachute training, to a night LCA attack by SAS troops on an island defended by HQ 3 Commando Brigade.

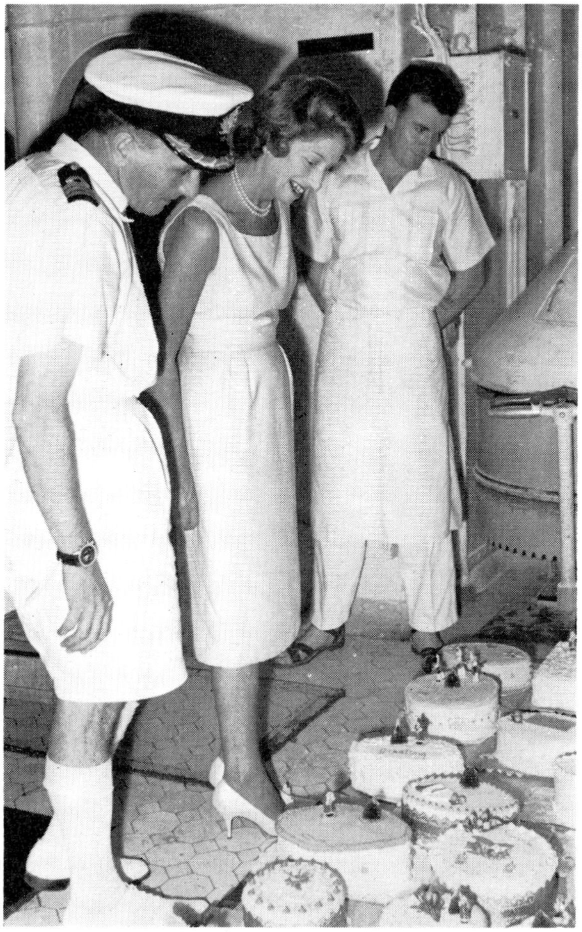

Christmas at Singapore—the cakes are admired.

On 17 January a helicopter of 848 Squadron experienced clutch failure and crashed in the bush near Muchea. The aircraft was one of eight which had been detached to RAAF *Pearce* for the training of Special Air Service troops. Fortunately the only passenger was the RM Special Boat Section Colour-Sergeant R. Adamson, and neither he nor the pilot Lt Cdr K. Mitchell was hurt.

Once the ship left Fremantle it was to be nearly 5,000 miles before her next port of call. On the morning of 19 February *Bulwark* anchored off the small island of

Australia

Perth 16.1.1962—Whirlwinds fly past . . .

. . . whilst the men march past.

A rare sight . . . stokers in smart white uniforms.

Preparing "The Bulwark Belle" for the children's party.

Socotra, which lies 150 miles north-east of the Somali peninsula and about 700 miles east of Aden. Here H.E. The Governor of Aden and the Commander-in-Chief, Middle East, visited the ship.

After four weeks of exercises in the Persian Gulf, and a three-day visit to Bahrein, *Bulwark* arrived back in Singapore on 23 March for a three-week maintenance period. During this period, the ship received a visit from the Minister of Defence, the Rt Hon. Mr Harold Watkinson, MP.

On 16 April it wasback to sea again. During the passage to Hong Kong the flight deck crews had an opportunity to familiarise themselves with the new Wessex helicopters, when a number of them from *Ark Royal*'s 815 A/S Squadron were embarked for training. Two weeks were spent in Hong Kong, when flights round the island were offered to prominent civilians of the colony. In place of floodlighting, *Bulwark*'s illuminations consisted of a dragon breathing fire, surmounted at the masthead by a flashing sign in Chinese characters which meant 'Hong Kong is a Happy Place'.

During the stay *Bulwark* was honoured to receive the Chief of the Air Staff, Marshal of the Royal Air Force, Sir Thomas Pike, GCB, CBE, DSO, and also His Excellency The Governor of Hong Kong, Sir Robert Black, GCMG.

One other event of note was a Search and Rescue Operation for the passengers of a junk which was believed to have capsized some fifty miles to the east of Hong Kong. Eight unsuccessful sorties were flown from sunrise on Sunday 29 April until the search was called off shortly before noon.

Soon after sailing from Hong Kong on Monday 7 May Exercise 'Common Assault 3' took place in the New Territories. Embarked for the occasion were the Commander British Forces, Lieutenant General R.H. Hewetson, CB, CBE, DSO, and the Commodore, Hong Kong, whose broad pennant *Bulwark* wore. The exercise involved the tactical landing of troops by helicopter and was their initiation into ship/shore movement.

On completion the ship sailed for Sandakan, where she was to rendezvous with 104 Mine-Sweeping Squadron of CMS's to carry out anti-pirate patrols. Police officers were embarked to assist in the identification of craft. The force was split into two, their mission being to neutralise the notoriety of the raiders by a show of force and to show goodwill in the remote islands and villages of North Borneo.

Once the ship arrived back at Singapore it was time for a short refit which meant that once again the ship's company had to be accommodated in HMS *Terror*.

During the middle of July, with the refit completed, *Bulwark* embarked officers and men of the Queen's Own Highlanders for exercises, working up towards the forthcoming SEATO Exercise 'Fotex 62'.

A final visit of the commission to Hong Kong ended on 31 August, when she set sail for Singapore and the

Bugis Street, Singapore—a favourite haunt.

Exercise "Duffle"—an unopposed landing.

Admiral's Annual Inspection. *Bulwark* left there on 12 September with the destroyer *Carysfort* in company, to be later joined by the frigate *Eastbourne*. Combined exercises were carried out whilst on passage to the Persian Gulf. *Eastbourne* was detached to Bahrein on the 22nd, but *Bulwark* and *Carysfort* anchored about thirty miles north of Muscat. Three days later they joined forces with the Amphibious Warfare Squadron assembled in the lee of Sir Abu Nuair Island for Exercise 'Duffel Two'.

During a visit to Kuwait, from 4-8 October, Captain Dalglish was received by the ruler, His Highness The Amir. Calls were also exchanged with H.B.M. Ambassador; the Kuwait Oil Company general manager; H.E. The Governor of Ahmadi; and the Acting Commander-in-Chief Kuwait Army (brother of the Governor). 40 Commando and 848 Squadron took the opportunity of operating with the Kuwaiti Army.

With the destroyer *Barrosa* in company, *Bulwark* left Aden on Monday 15 October. Later joined by the frigate *Brighton*, the ships anchored off Port Victoria, Mahe in the Seychelles on the 19th. To publicise their arrival *Bulwark*'s helicopters made a 'round the island' flight—one towed beneath it a union flag and another a white ensign.

On Trafalgar Day—a Sunday—a strong party landed for Divine Service in the Anglican Cathedral, which greatly boosted the morale of the Dean and the Anglicans in this mainly French and Roman Catholic community. The ship was also able to parade a guard and band for 'Colours' for the first time this commission.

Leaving the Seychelles on 22 October, the ship headed for Mombasa where she secured alongside with *Brighton* nearby. After 24 hours alongside, the ship shifted berth to the stream, with her stern to a buoy and two anchors down. She then reverted to eight hours notice for steam to carry out self-maintenance. Many of the ship's company took the opportunity of all night and station leave, and over 200 visited the Royal Tsavo National Park.

A detachment of four helicopters was based for four days at Voi, their job being to assist the National Park game warden in counting elephants and plotting their movements.

Attempts by two teams were made on Mounts Kilimanjaro and Kenya. Despite mountain sickness, both teams got above 16,000 feet but neither was able to complete their climb.

Bulwark returned to Aden on 20 November. There, the ship's company had the great pleasure of witnessing the arrival of HMS *Albion*, who had come to relieve them. Five hectic days were spent transferring stores and 40 Commando to *Albion* before commencing the last stages of the voyage home. A swift passage through the Mediterranean and a brief visit to Gibraltar enabled *Bulwark* to arrive in Devonport on 17 December, 1962.

Kilindini Road, Mombasa—a good spot for buying presents on the way home.

6

Fifth Commission

1963 to 1965

Most of 1963 was spent under refit in Devonport Dockyard. In addition to the usual servicing and repair of equipment, the facilities for handling aircraft were improved. The markings on the flight deck were adjusted to prepare for the larger Wessex helicopters to be carried.

In September, Captain P.J. Morgan was appointed to the ship but the formal commissioning ceremony was not held until 3 December, 1963.

The ground crews of 706 Bulwark Flight of Wessex helicopters, men of 43 Commando Royal Marines and a battery from 29 Commando Regiment, Royal Artillery joined the ship during the last week of February, 1964.

Twenty six days after leaving Plymouth, *Bulwark* arrived in Singapore (on 2 April).

OPERATIONS IN BORNEO

The troubled situation in Sarawak had necessitated the permanent detachment ashore of half of 845 Squadron. Although the other half remained on board at first, each group changed places every three months. Their task was to give the security forces the speed and mobility to enable them to combat infiltrators throughout the great length of the jungle border area.

Exercise "Sandfly II"—A short break in the Mediterranean.

Before the ship left Devonport on 7 March, she received a visit from Earl Jellicoe, First Lord of the Admiralty.

Operational requirements prevented the ship calling at Gibraltar, and only one day was spent in Malta—without leave—before embarking on a two-day Exercise 'Sandfly II' off the North African coast. The ship's LCAs suffered unfortunate mishaps and on completion of the exercise were left grounded on the beaches, to be recovered later. 43 Commando were not re-embarked but were airlifted back to the UK.

After a quick passage through the Eastern Mediterranean and the Suez Canal, *Bulwark* arrived at Aden on 23 March where she was to relieve *Albion.* Two days were spent transferring stores and equipment, including *Albion*'s LCAs. *Bulwark* also received *Albion*'s Squadron of six Wessex helicopters. Combined with 706 B Flight they now became 845 Naval Air Squadron.

Aden—take over from **Albion.**

En route to Singapore, with 845 Squadron Wessex embarked. Two R.A.F. Whirlwinds, one Belvedere, and a lone R.N. Gannet also take passage on board.

Singapore—the usual berth in the Stores Basin.

Air Ops in Borneo

845 Squadron detach to Sibu.

Not exactly Heathrow—but almost as busy—Nanga Gaat, forward operating base.

The Borneo half of the Squadron was in three detachments—Sibu, Nanga Gaat and Simanggang—with just eight Wessex and a Hiller to support the entire 2nd and 3rd Divisions of Sarawak. The 3rd Division is a vast area, similar in size to Wales, mostly mountainous and almost entirely covered in jungle. Sibu, the Capital, was a large town of mainly Chinese population.

The Squadron's main base was cramped bungalows near the airport, some three miles from Sibu town. This detachment supported all military activity in the western half of the 3rd Division.

The Simanggang detachment was withdrawn in June 1964 and Nanga Gaat remained the Squadron's permanent forward base deep in the eastern half of the 3rd Division. This detachment developed a special relationship with the SAS, the Guards Independent Parachute Company, the Gurkha Company and the Independent Company of the 2nd Battalion Parachute Regiment.

Although no positive enemy contacts were made in the 3rd Division of Sarawak, the Squadron's helicopters enabled one Battalion and the Special Parachute Companies to keep control over 25,000 square miles against the triple threat of border crossing, sea incursion and internal subversion.

As one would expect, the Squadron operating in the Commando role over Borneo jungle had various aircraft incidents culminating in a series of major accidents between February and April 1965. These resulted in the loss

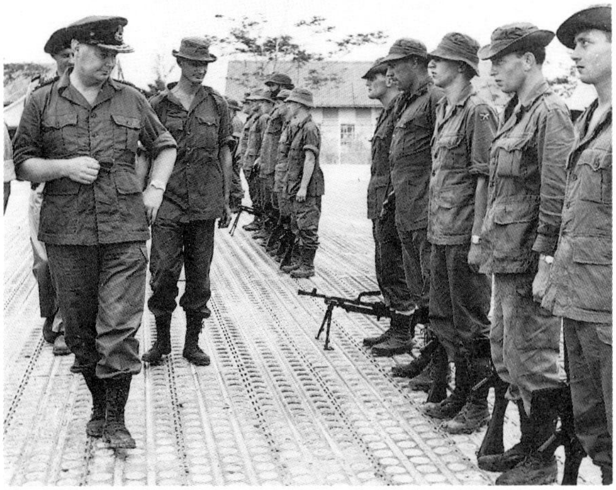

Why should Britain tremble? The Air Officer Commanding, Far East, inspects the "Junglewood Fighters" of 845 Squadron.

It was never like this at Culdrose!

Whoops!—But no one was hurt.

of 5 aircraft, 3 pilots, 2 aircrewmen and 11 soldiers. However, in spite of these disasters, the entire Squadron reacted with magnificent resilience. Their efforts resulted in the award of the Boyd Trophy—presented annually for the finest feat of Naval Aviation. The following is the citation of the Flag Officer Naval Air Command, Vice Admiral Sir Richard Smeeton:-

> "By their operations and exercises, both ashore and afloat, 845 Squadron set a standard of efficiency in Borneo unsurpassed by any other Squadron deployed there and made a considerable contribution to the defence of Malaysia. They earned the praise of all. Their flying was of the highest order and the enthusiasm and professional ability of the maintenance ratings enabled the Squadron to achieve an outstandingly high serviceability rate, and the distinction of meeting every demand made upon it."

The Squadron's Commanding Officer, Lieutenant Commander G.J. Sherman, was awarded the Order of the Star of Sarawak and appointed a Member of the British Empire.

During April 1964 *Bulwark* made several voyages to the coast of Sarawak, supporting the shorebased half of 845 Squadron. The ship did not anchor but continued to steam up and down the coast, and 'darkened ship' each night because of the war situation in the area.

In the first week of June the ship joined other ships of the RN and USN for Exercise 'Ligtas' in the area of the Philippines. On completion, it was back to Sarawak to re-

A terrorist is flown in for interrogation.

Singapore and a change of screw believed damaged by a semi-submerged tree.

plenish 845 Squadron, prior to a 10-day routine maintenance period in Hong Kong.

Exercise 'Horses Neck' began on 17 August off the east coast of Malaysia. However, two days after 42 Commando had been put ashore, the helicopters were withdrawn from the exercise because of a reported 'incursion' by Indonesian troops in the Pontian area of Malaysia.

A few weeks later a large number of Indonesian troops landed by parachute in the Labis area. Four Wessex and a Hiller aircraft were sent to assist the UK and Commonwealth troops.

On 14 November *Bulwark* arrived in Hong Kong for another visit, during which an officer and eight ratings set off in a boat to land on a nearby island and climb a mountain. However, by an error in navigation, they finished up in Communist Chinese waters. They were stopped by rifle and grenade fire, and taken into custody for three days.

After a brief trip to Tawau, to embark the helicopters and men of 845 Squadron whose tour of duty in Borneo was completed, it was back again to spend Christmas and the New Year in Hong Kong.

There the command of the ship was handed over by Captain Morgan to Captain D.B. Law, CBE, DSC, RN.

Leaving Hong Kong on 5 January 1965 it was back to Sibu, followed by a four day visit to Manila, and thence to Singapore where the ship remained for the whole of February.

Exercise 'March Hare' was carried out between 2 and 5 March. It was one of the few times that a Commando had been landed ashore, supported and withdrawn, with all activities taking place during the hours of darkness. On completion, it was back to Singapore for another two weeks.

'Fotex 65', an exercise of convoy work, anti-submarine and air defence duties, took place during the last two weeks of March off the west coast of Malaysia.

Forty ships of the RN, RAN, RNZN and the Royal Malaysian Navy took part in these activities in the Singapore practice area and in the Malacca Straits. An impressive array of 'flat-tops'—*Melbourne, Eagle, Victorious* and *Bulwark*—was included.

On 5 April *Bulwark* entered Singapore Dockyard for a seven week maintenance period. As this was to entail six weeks in dry dock, the ship's company moved ashore into HMS *Terror.*

By late May the end of the commission was in sight. The advance party had been flown home and their reliefs were

"We're the most savage, bloodthirsty, treacherous tribe in Borneo—we're the 845 Squadron Royal Navy!"

An RAF Shackleton comes close—very close—to have a look.

settling in. After Exercise 'Windy Weather' in June the ship returned to Borneo to pick up the members of 845 Squadron, who by this time had been relieved ashore by the men and the new Mk 5 Wessex helicopters of 848 Squadron. They had been brought out by *Albion.*

Then followed what many considered the highlight of the commission—a visit to Western Australia. After an enjoyable stay in Perth and Fremantle, *Bulwark* set sail for home. She called at Aden but the passage through the Mediterranean was made without stopping and she arrived in Plymouth on 3 September 1965.

During the commission *Bulwark* had steamed 83,729 nautical miles—the busiest month being March 1964 when 7,902 nautical miles were covered.

7

Sixth Commission
1966 to 1967

On her return from the Far East in September 1965, *Bulwark* was taken in hand for refit by Devonport Dockyard. In addition to the normal servicing and equipment repairs, extensions were made to the ship's flying control facilities.

The commissioning ceremony took place on 22 April 1966, and eight days later she sailed for the first of her trials and a shakedown. On completion of her Group 2 Trials on 26 May she visited Portsmouth.

A rare bird . . . A Kestrel, would you believe?

On 18 June the ship carried out trials with a small aircraft, the Hawker P1127, or Kestrel. At the time it attracted little attention, but it introduced the principle of vertical/short take-off and landing (V/STOL) at sea. The aircraft was to enter service some years later in the guise of the Hawker Siddeley Harrier.

At Plymouth *Bulwark* embarked her Air Squadron and 42 Commando, RM, who were to take part in Exercise 'Dry Fly' on the west coast of Scotland. The ship's Squadron was again No. 845, but it had converted to the Mk 5 Wessex helicopters. The Mk 1 helicopter had only had a single engine but the Mk 5 had two centrally coupled engines which not only gave greater power, but also improved safety, for if one engine failed, the helicopter could still fly using the other.

The ship's company, Air Squadron and Commando were able to enjoy some leave before *Bulwark* sailed from Plymouth on 13 August. Once the ship cleared harbour, the helicopters were embarked. An advance party of air crews and maintenance staff had been flown out to Sarawak to relieve the part of HMS *Albion*'s 848 Squadron who were based ashore and allow them to rejoin their ship.

After a few days in Gibraltar, the ship made her way across the Mediterranean to arrive at Port Said on 28 August. Off the southern end of the Suez Canal *Bulwark* rendezvoused with *Albion* so that some stores could be transferred.

Before crossing the Indian Ocean, Exercise 'Michael' was carried out on the Aden Peninsula. Whilst on passage for Singapore, *Bulwark* stopped at Gan, a small island in the Maldive group, to supply the RAF airfield there with motor transport fuel.

Two weeks were spent in Singapore before sailing on 5 October to call at Labuan Island, off the north-west coast of Sarawak, where the men and equipment of 845 Squadron's advance party were waiting to embark. The situation in Malaysia had now reverted to normal and there was no need for a Naval Air Squadron to be based ashore. Thus, for the first time in nearly four years, the complete Squadron was able to be embarked.

On 18 October 1966 *Bulwark* anchored in Shoalwater Bay in northern Queensland. There, Exercise 'Barrawinga' was carried out. The 'friendly' troops were men of the Royal Australian Regiment and the enemy were 42 Commando. For the purpose of the exercise, 845 Squadron was split into two flights—one flight remained embarked, while the other was based on a disused airstrip in the middle of the bush, about thirty miles from Rockhampton. An accident to a helicopter resulted in all the shore-based Wessex aircraft being grounded, so the exercise was completed using the ship's LCAs.

Sixteen days were spent visiting the capital of Queensland, Brisbane, where several thousand of the local

Floodlit for Christmas and New Year at Hong Kong.

Trengganu Flood Relief . . .

population took the opportunity of looking over the ship.

On her way back to Singapore, *Bulwark* called at the Sarawak capital of Kuching to pick up men and helicopters of the RAF and ferry them back to Singapore.

Exercises off the Malaysian coast, with men of the 2nd/6th Gurkha Rifles embarked, were followed by a visit to Hong Kong. The Christmas and New Year period was spent there during which time Captain A.M. Power, MBE, RN, assumed command of *Bulwark*.

Under her new command, *Bulwark* left Hong Kong on 3 January. Two days later an SOS was received from the Yugoslavian steamer *Kostrena*, who had gone aground off the Paracel Islands in the South China Sea after her engines had failed. *Bulwark* stood by the steamer until a tug arrived from Hong Kong, after which she continued on to Singapore.

During January 1967 serious flooding occurred in the Malaysian State of Trengganu. The ship's landing party and helicopters of 845 Squadron were sent to the area to help with the distribution of food.

The next six months were spent on various exercises with her own Commando and other regiments, interspersed with periods in Singapore and Hong Kong.

The ship's visit to Hong Kong in June was considered by the official Chinese Communist Party organ "People's Daily", an intimidation of the Chinese compatriots in Hong Kong. It said Britain was vainly trying to repeat 19th century gunboat diplomacy. The military exercise carried out by her Royal Marines was described as "a brazen provocation at the very gate of South China".

However, *Bulwark*'s presence might not have pleased

Amphibious exercise with **Fearless**, *March 1967.*

Stranger on deck—fixed wings return—from the Army Air Corps.

Off to Aden

The Commando, and all their equipment, embark.

The start of a long passage to Aden.

Safely back to Singapore . . . The commission ends and the entire ship's company prepare to fly home for some much awaited leave.

the Reds, but she was a welcome sight to the people in Hong Kong.

The capability of the Far East Amphibious Forces Group to land and sustain a Brigade Group was demonstrated by an exercise held on the east coast of Malaysia involving the 3rd Commando Brigade. The group sailed from Singapore in *Bulwark*, assault ship *Fearless* and the Army LSTs *Gannet* and *Kittywake.* Two thousand men were landed near the town of Kuantan by helicopter and landing craft.

In July *Bulwark* operated off the main Royal Thai Marine Corps base of Sattahip on the west coast of the Gulf of Siam. The opportunity was taken to give the Corps their first helicopter training. Lectures and demonstrations on shore were followed by the embarkation of the Thai Marines in *Bulwark.* After leaving Sattahip, the ship anchored off the holiday resort of Bangsaen and some of the ship's company were able to visit Bangkok, which was about two hours coach ride away.

At the end of August *Bulwark* once again entered dry dock in Singapore. On 28 September the ship sailed for a post-refit shakedown before heading eastward across the Indian Ocean to Aden.

The sixth commission ended in Singapore on 23 October, 1967.

8

Seventh Commission

1967 to 1969

No sooner had the ship re-commissioned, than she returned to Aden . . . Here, Sea Vixens from **Hermes** *show off as they fly past.*

Another use for the flight deck. 95 Battery, Royal Artillery keep their hand in.

Leaving Singapore—homeward bound.

For the second time *Bulwark* was recommissioned in Singapore. This enabled her, and *Albion,* to be in the Aden area to cover the withdrawal of the UK Forces. *Bulwark* embarked 45 Commando—and as that Commando had been in Aden for seven years, their equipment filled all the available space in the ship. She returned them to Singapore, from where 45 Commando were airlifted back to UK in preparation for their new role, Arctic Warfare.

Christmas and the New Year was spent in Singapore but soon afterwards exercises were carried out in the Indian Ocean with other units of the Far East Fleet.

Lieutenant William Wood made the 10,000th aircraft

The calm of Durban after the storms of the Indian Ocean.

Portsmouth at last

Only another half an hour before families are re-united.

landing in *Bulwark* since the ship commissioned at Devonport in April 1966, when he landed a Wessex Mk 5 helicopter of 845 Squadron. In addition to Royal Navy Wessex and Wasp helicopters, Beaver aircraft of the Royal Corps of Transport, Whirlwinds, Wessex and Belvederes of the RAF, and Sioux of the Royal Marines helped to make up the 10,000 landings in 15 months.

In March 1968 *Bulwark* left Singapore for the long journey home. The monotony was broken by a visit to Durban in South Africa. On passage there, the ship encountered heavy seas and strong winds, and her arrival—on 6 April 1968—was 24 hours behind schedule. Hundreds of people took advantage of it being a Saturday to line 'N' Berth at the docks to greet her arrival. As *Bulwark* made her way through the narrows, the Lady in White, Mrs Perla Siedle Gibson, sang many of the old wartime songs—her words clearly heard by those on board.

Captain Power revealed at a Press Conference held on board shortly after berthing, that he had been promoted to Rear Admiral and Durban would be the last foreign port he called at before taking up his shore appointment at Portsmouth.

Cars packed the approaches to the docks, whilst thousands of sightseers waited patiently in a queue, sometimes 300 yards long, to go on board.

Captain Power and his officers were entertained at a civic luncheon, and they reciprocated by being hosts at a cocktail party attended by guests from all over Natal.

Bulwark was given a rousing farewell when she sailed from Durban on the 9th. On passage up the English Channel, 845 Squadron of Wessex Mk 5 helicopters were flown off to the Naval Air Station at Culdrose in Cornwall, and the ship arrived in Portsmouth on 25 April.

When the ship's company returned from leave, they found that their new Commanding Officer, Captain J.A. Templeton-Cotill, RN, had been appointed to the ship on 1 May 1968, and that *Bulwark*'s new role was to be part of the NATO Fleet.

Exercise 'Polar Express', designed to test the deployment of the Allied Command Europe Mobile Force concept, began off the northern coast of Norway on 3 June. In addition to the land and air components of the Force, a number of HM ships with ships from other NATO countries, took part. In particular the exercise provided a good opportunity to evaluate the problems of operating a Commando Carrier, with an embarked commando group, on the northern flank of NATO. 45 Royal Marine Commando, embarked in *Bulwark*, acted as the enemy force in conjunction with units of the Norwegian Army.

The few weeks *Bulwark* had spent in the UK since her return had not permitted her helicopters to be modified for operations in Arctic conditions. However, very low temperatures were not anticipated at that time of year and were not, in any case, considered likely to limit the object of the exercise.

In the event, the first 17 hours of the exercise took place in freezing conditions, and the initial landing of the Commando had therefore to be carried out using *Bulwark*'s landing craft. After an improvement in the weather, helicopter operations in support of the landings were able to start and run round the clock. Much valuable experience was gained by both the Commandos and the Norwegians, and the exercise again demonstrated the versatility of the Commando Carrier.

During the last week of July 1968, 50 ships of the Western Fleet, including *Bulwark*, assembled at Rosyth. The aim of the assembly was to allow Flag and Commanding Officers to meet and discuss current problems; to hold Fleet Examinations; and carry out harbour training. Amongst the demonstrations arranged was a Commando attack by a force from *Bulwark*.

After a visit to Rotterdam, the ship returned to Portsmouth in August. On arrival the ship's company trans-

Grounded—the snow and ice win.

The Commando get ashore —eventually.

The show goes on.

ferred to her sister ship *Centaur,* which was to act as an accommodation ship whilst their own ship underwent a seven-month refit.

Early in March 1969 the ship's company returned to their own vessel, and on 17 April she sailed from Portsmouth, arriving in Plymouth the next day. There a procession of lorries trundled into Devonport Dockyard throughout the day as Royal Marines of 41 Commando boarded the ship to start a four month tour of duty in the Mediterranean. Other commandos to embark belonged to 7 Battery of 95 Commando Light Regiment, Royal Artillery, based at the Royal Citadel, Plymouth.

Work-up trials were conducted whilst the ship was on passage to the Mediterranean. After a brief stop at Gibraltar she proceeded to Cyprus, where she arrived on 26 April. There, 845 Squadron, equipped with Wessex Mk 5 helicopters, and 41 Commando were flown ashore to conduct exercises with the UK Forces based on the island. *Bulwark* returned to Malta for a ten-day self-maintenance period.

On completion of the Exercise 'Grecian Vase', she made her way to the coast of Salonika for an Exercise, codenamed 'Olympic Express', in company with ships of five other nations.

In the middle of June the ship headed westwards through the Mediterranean. On passage she conducted various inter-ship evolutions and a number of replenishment-at-sea exercises with various RFAs. After a somewhat leisurely passage, she arrived at the French naval port of Toulon for a two-week maintenance period.

The visit to Toulon was noteworthy and unusual in that it was the first time for many years that a ship of the *Bulwark*'s size had undertaken a self-maintenance period in a foreign port. In helping to promote the 'twinning' of the cities of Portsmouth and Toulon, the opportunity was taken of presenting gifts on behalf of the City of Portsmouth. These gifts, which included the Portsmouth Guildhall Flag and volumes of the Portsmouth official records dating back to 1835, were presented to the Maire of Toulon at an impressive ceremony held in the Municipal Buildings on 24 June. Many French civic officials and senior naval officers were present. A Vin D'Honneur followed the presentation to complete a very friendly and happy occasion.

Visits were also made to Venice and Villefranche. *Bulwark* was the largest warship ever to anchor opposite the famous St Mark's Square in Venice and so many people wanted to come on board that the ship was open to visitors on three successive days.

At Villefranche the ship was honoured by a visit from Prince Ranier and Princess Grace of Monaco, accompanied by their daughter, Princess Caroline. Their Serene Highnesses toured the ship for well over an hour, talking to many officers and ratings before lunching with the Captain.

Leaving Villefranche, homeward bound, *Bulwark* made a brief stop at Gibraltar. There she took part in a joint Services Exercise 'Rock Climb', in which, as part of their normal training, the RMs carried out a helicopter borne assault on The Rock. The defending forces included the 2nd Battalion Royal Irish Rangers, the Royal Anglian Regiment and the Royal Engineers.

The Air Squadron was disembarked whilst the ship was off Cornwall and the Commando landed at Plymouth, before the ship arrived back in Portsmouth on 15 August, almost four months to the day since she had left.

After summer leave and Portsmouth Navy Days, *Bulwark* returned to the Mediterranean.

During October, four enjoyable days were spent at Genoa. The Italian harbour authorities arranged for the ship to be alongside in the excellent Andrea Doria berth, where she could be clearly seen from a wide area of the city. For two successive days, the ship was open to visitors and about 10,000 came on board. On 15 October, 34 officers and men were received in audience by His Holiness the Pope. *Bulwark* then returned home to pay off and prepare for her next commission.

9

Eighth Commission

1969 to 1972

Bulwark, now to be commanded by Captain B.H. Notley, had a 'high powered' ecclesiastical send off to her eighth commission on 10 November, 1969. The Chaplain of the Fleet and the Principal Chaplains of the Church of Scotland, the Free Churches and the Roman Catholic Church all took part in the recommissioning ceremony at Devonport.

Working-up exercises were carried out during November and December, but the ship returned to Devonport for Christmas leave. The last goodbyes to UK were said at Portsmouth when *Bulwark* left there early on 27 January 1970, and headed down Channel. 848 Air Squadron was embarked at Plymouth; then the ship visited Gibraltar briefly, but instead of turning in to the Mediterranean, she continued down the west coast of Africa.

On the way to Capetown strenuous efforts were made to clean the ship—both inside and out. However, the trip was not all work. There was some 'fun in the sun', including the usual sporting activities on the Flight Deck. 'Crossing the line' was celebrated, and 'Hands to Bathe' was being enjoyed until the cry 'Shark!'—Olympic records were then broken by nearly everyone in the water.

Following five days in Capetown, the ship sailed on eastwards. Cyclone 'Jane' was encountered while on passage to Gan.

The entry into Singapore was made the more impressive by the mass departure of 847 and 848 Squadrons for Simbang. A month was spent on self-maintenance before leaving for Hong Kong. The Sultan of Brunei visited the ship arriving by FPB and departing by helicopter while the ship was off his coast. The stay at Hong Kong was brief and *Bulwark* was soon heading for the highlight of the Far East tour. Between 23-28 April the ship visited Kobe in Japan during which time many men were able to visit the International Trade Fair—Expo 70—at Osaka. It was often difficult to get ashore because of the thousands of visitors wanting to come on board each day.

A self-maintenance period at Hong Kong and a dockyard assisted maintenance period at Singapore, were followed by more exercises. The best day's condensed flying was a display given by 847 and 848 Squadrons for the West German press and television. It blew half a gale throughout the display and, on and off, rained in torrents. Despite this 'the show went on'.

A relaxing "Hands to bathe"...

... until someone shouted "shark!"

Between 13 June and the end of the month, combined forces of Australia, Malaysia, New Zealand, Singapore and the United Kingdom, involving over 20,000 men, some 50 ships and nearly 200 aircraft, carried out the Joint Maritime and Air Defence Exercise which formed the main phase of Exercise 'Bersatu Padu' (a Malayan phrase meaning 'complete unity'). Its aim was to operate

Below decks the work goes on—boilers are cleaned . . .

under conditions which would prevail after the United Kingdom's withdrawal from the area; and to practise setting up the necessary organisation to introduce, support and train forces from outside the area. It opened with an amphibious landing on the east coast of Malaysia, followed at the end of the main phase by an air defence

. . . and technical information recorded.

A colourful entry was made to Kobe.

exercise over Butterworth, Singapore and the fleet at sea.

At the beginning of the exercise, 40 Commando RM from *Bulwark* took part in the amphibious landing, quickly capturing the airstrip at Penarek. Later the Commando moved into the Marang area, Western Malaysia, for further action against 'enemy' strongholds.

At sea the Allied Naval Task Force of 15 ships including *Bulwark* and the Australian carrier *Melbourne* were frequently atacked by 'enemy' bombers and the Fleet's air defences were tested severely.

After off-loading troops and stores, *Bulwark* left for Djakarta on the start of the long journey home. Two days in Djakarta—during which time 40 senior Indonesian officers and VIPs came on board to witness a demonstration of the ship's capabilities—and the ship was off again to Fremantle in Australia where she was berthed from 14 to 20 July. As the ship entered harbour, 10 Wessex Mk 5 helicopters of 848 Squadron were launched and, led by a Wasp helicopter, carried out a fly past over Perth and Fremantle, landing back on board after the ship had secured alongside.

It was difficult to get ashore because of the thousands of visitors each day.

En route to the UK the ship called at Durban and Gibraltar. She arrived at Durban on 31 July—the last day for the general issue of rum in the Royal Navy. To mark the occasion the ship's company presented a rum barrel to the local branch of the Royal Naval Association.

Bulwark arrived at Plymouth on 20 August, to enter the dockyard at Devonport for refit; so the ship's company moved ashore into the barracks, HMS *Drake*, for the next six months.

The ship—with the horrors of the refit behind her—left the dockyard on 9 March 1971 for trials. Three Wessex Mk 5 helicopters were embarked, one of which was used for a special aircraft trial and the other two for mail, stores and passenger trips. She returned to the dockyard on 25 March for a further rectification period, but 16 April saw the ship sailing for the work-up and shakedown period.

Leaving Spithead on Thursday 13 May *Bulwark*, in company with the frigate *Andromeda*, headed eastwards to the Baltic for a visit to Stockholm. On the 17th, the Commander-in-Chief, Western Fleet, Admiral Sir William O'Brien, and his staff, arrived on board in a Swedish Boeing Vertol 107.

The ships were dressed overall on the 19th, to greet the heir to the Swedish throne, Crown Prince Carl Gustav, who toured *Bulwark*. He was returning the call made on his grandfather, King Gustav VI Adolf, by Admiral O'Brien.

Many of the ships' companies took the opportunity of visiting the preserved hull of the ship *Vasa*, which had

Durban, but a sad day for the Royal Navy: the last issue of rum, 31 July, 1970.

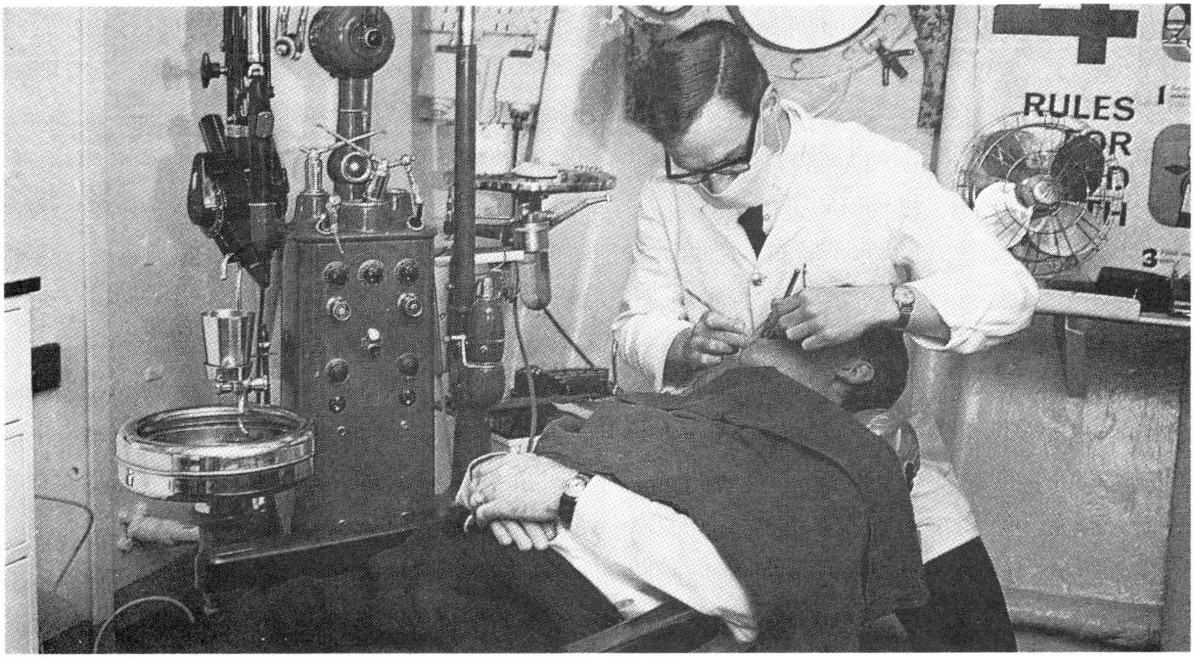

The ship's dentist carries out "planned maintenance and essential defects", as the ship heads for her own refit.

capsized 333 years previously and had been salvaged after over 300 years under water.

After four days in Stockholm, *Bulwark* and *Andromeda* made a night dash to Helsinki for another enjoyable visit. The ships were berthed in South Harbour at Katajanokka.

On arrival, the Commander-in-Chief and the two Commanding Officers made several official calls, including signing the visitors' book in the Presidential Palace. On the Sunday afternoon wreaths were laid on the Cross of War Heroes and on Marshal Mannerheim's grave at Hietaniemi.

Arriving at Stockholm.

On her return to Plymouth, *Bulwark* met sea fog, so instead of going alongside to embark 41 Commando, she was only able to creep to 'C' buoy in the Sound.

Exercise 'Dry Fly III' commenced on 12 June with *Bulwark* anchored in Loch Fyne, some 25 miles from the exercise area. It involved the landing, support and withdrawal of 41 Commando in the west of Scotland. A forward air base was set up at Inverary to which half of 845 Squadron 'emigrated' to live in tents.

A week's self-maintenance at Devonport followed. During this period there was a sudden requirement to prepare the Flight Deck to receive aircraft for refuelling. They had been searching for survivors of a disastrous Plymouth-Fowey yacht race. The weather had 'clamped down' at Culdrose, so *Bulwark* hosted four Sea Kings and two Wessex Mk 5s together with their crews for the night.

Bulwark left Devonport on 30 June for a five day visit to Liverpool. On the ship's return, a 'families day' was held during which three Wessex and a Wasp gave a mini display.

Six weeks were spent maintaining, with dockyard assistance, then 41 Commando RM, and supporting gunners from the Army's 29 Commando Light Regiment, Royal Artillery, from the Royal Citadel, Plymouth, were embarked and the ship headed for Gibraltar for a two day visit. On leaving the Rock, she turned east and her transit of the Mediterranean was watched by Russian ships. On arrival at Larnaca Bay, Cyprus, the Commando were disembarked, and the next day the Air Squadron flew ashore to RAF Akrotiri as the ship departed for Istanbul.

For this visit *Bulwark* was accompanied by the RFAs *Tidereach* and *Retainer.* Up to 250 guests attended an official reception on board *Bulwark;* a ship's company dance, hosted by the British Community Council, was held at the old British Summer Embassy at Tarabya; and many sightseeing tours were arranged.

After five days *Bulwark* slid out of the Bosphorus, through the Dardanelles and back into the warmer waters of the Mediterranean. The ship returned to Cyprus to pick up the Squadron and Commando in preparation for Exercise 'Royal Return'.

More than 1,000 Royal Marine Commandos, embarked in *Bulwark* and *Fearless,* spearheaded Britain's contribution to the NATO Exercise 'Deep Furrow 71' which took place in the eastern Mediterranean in October. For this exercise the NATO Fleet, which included other British ships and units of the US, Turkish, Greek and Italian Navies, gathered at Soudha Bay, Crete. After an opposed transit of the Aegean, Royal Marines from *Bulwark* and *Fearless,* and US Marines, landed in the Turkish Thrace to assist Turkish forces. Among those attending the land demonstration was the First Sea Lord, Admiral Sir Michael Pollock, who later visited *Bulwark, Fearless* and the destroyer *Norfolk.* On completion of the exercise, the RN ships made another operational visit to Istanbul.

Bulwark then made a tour of the Adriatic before heading for home.

While the ship was on passage from Venice to Trieste, she encountered a hurricane, which prevented her from going alongside at Trieste, so she had to spend the next three days riding out the storm. *Bulwark* left Trieste 24 hours earlier than planned because of the bad weather and had to leave 120 of the ship's company ashore to follow in an escorting RFA.

After brief calls at Malta and Gibraltar, and having recovered her 'lost' crew members, the ship made all speed to Devonport, where she arrived on 8 December 1971.

The following week Captain L.R. Bell Davies assumed command of the ship.

On arrival at Devonport it was revealed that two weeks earlier, on 26 November, while the ship was at anchor off Trieste, fire had ravaged 'B' boiler room. For their heroic actions during this drama, British Empire Medals for Gallantry were awarded to Acting PO MEM Joseph Thompson, of Newcastle, and MEM 1 Peter Simcock, of Cheshire.

The citation said of PO Thompson: "Without his resolute leadership and clear thinking, it is doubtful whether the crew of B1 boiler room would have survived."

PO Thompson was on watch, in the middle of the night, when a boiler fuel hose burst. Hot oil sprayed the bilges and the boiler, and immediately caught fire.

The crew of five young men almost succeeded in quenching the flames, but the foam blanket was not quite complete when the fire burst out again. It burned fiercely on either side of the group and blocked their exit route.

The boiler room filled with smoke, the heat grew in intensity and, not unnaturally, the men—two of them under 18—became apprehensive. But PO Thompson kept calm and, as he retreated, shut off everything to stop the fire from spreading and to safeguard machinery. Then he mustered his men under a ventilation shaft near the foot of the escape tunnel.

A rating tried to open the escape hatch but it was jammed and he fell back. As the six waited for outside assistance they remained calm in the choking smoke and heat of the burning boiler, being trapped for 42 minutes.

When the alarm was raised, MEM Simcock got into a fire fighting suit. Others were driven back by the flames, heat and smoke but eventually MEM Simcock burst through the flames and reached the party. The escape hatch was still jammed and he had to battle back through the fire to get help. MEM Simcock, whilst in the blazing boiler room, suffered burns through his fire suit.

The ratings were rescued, some of them barely conscious, and the boiler room was 'steam drenched' to cut off the air supply to the fire. Damage was extensive and fittings were melted by the heat.

The second citation said: "Simcock, only 19 years old, spent 20 minutes almost alone in a compartment filled with heat, flames and smoke of a major fire and carried out the task he had been given . . . "The trapped men were recovered unharmed and it became possible to close down the compartment and fight the fire in time to stop it getting out of hand.

"MEM Simcock's courage, devotion to duty and coolness in the face of considerable danger were in the highest traditions of the Service."

No-one was hurt but the Devonport period was expected to be extended to allow the fire damage to be repaired.

The work, it was thought, would take eight weeks, but the ship was needed urgently in the Mediterranean and the work was completed in just 12 days. Bad weather delayed her departure for 24 hours, but *Bulwark* left Plymouth on 13 January 1972. She remained in the Malta area until the end of February, acting as HQ ship for 'Operation Exit'—the withdrawal of the RN from Malta.

Mr Dom Mintoff, Prime Minister of Malta, visited *Bulwark* shortly before signing a 7 year defence programme with Britain. He made a 20 minute flight above the Island in a helicopter, piloted by Lt Cdr J. Rawlins, CO of 845 Squadron.

Whilst in Malta *Bulwark* received a message that the Greek ship *Ermioni* had foundered 25 miles to the north of the Island. Within 45 minutes two Wessex aircraft were on the scene searching for survivors. A lifeboat was found with 14 men in it and these were airlifted to *Bulwark*.

Two and a half months were spent in Malta, during which time 1,000 deck landings took place. As *Bulwark* left Grand Harbour, 9 Wessex Mk 5 helicopters performed a close formation fly-past.

41 Commando and 848 Air Squadron were embarked for

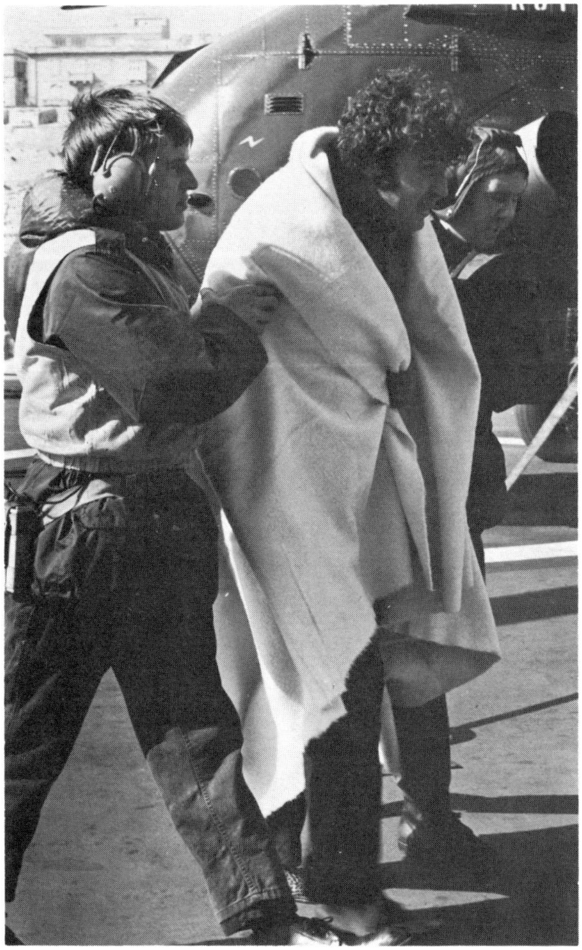

Saved—a survivor from the Greek ship **Ermioni** *is brought aboard.*

Maltese Interlude . . .

The ship spent 2½ months at this buoy whilst delicate negotiations were carried out regarding the future of British Forces in Malta.

Initially life was very busy

The ship's company loaded merchant ships . . . with private cars.

. . . and personal baggage.

A wide range of service equipment was stripped and returned to UK.

The saluting guns were removed from Fort St. Angelo.

Celebrating the 1,000th deck landing. All on the ship in harbour during "Operation Exit".

As time went on, the long days had to be filled

*Many rounds of "It's a Knock-Out" (***Bulwark*** style) were held on the flight deck—to the amazement of the Maltese ashore.*

The Captain (a submariner) took flying lessons—note his "L" Plate.

Others found more energetic pursuits ashore.

Visitors were few . . .

The Captain points out "The Cruel Sea" beyond the breakwater in Nicholas Monsarrat.

Mr Borg, dghajsa-man to visiting RN ships for a life time, is suitably rewarded as the ship prepares to sail .

Mr Mintoff, the Maltese Prime Minister, is given an aerial view of "His" Island.

Farewell Malta . . .

passage across the Atlantic to take part in Exercise 'Rum Punch' with US Forces. It took place round the island of Vieques, which lies to the east of Puerto Rico. Whilst in the area, *Bulwark* visited the US Navy's new base at Mayport in Florida, which had been dug out of the virgin sand by the USN.

On completion of the exercise, *Bulwark* returned to Malta to disembark 41 Commando. Once they were ashore, the ship sailed for the UK, calling at Plymouth before going on to Portsmouth where she arrived on 17 June 1972 to begin a 'Docking and Essential Defects' period.

Full speed for . . .

. . . Portsmouth.

Off to the States

77 *A drink on the way*

Arrival at Mayport . . .

. . . which had its attractions.

On 17 July *Bulwark* joined other units of the Fleet at Devonport, for the week long Exercise 'West Hoe'. During August she set off for a work-up cruise round the Scottish coast. Then she returned to Plymouth for Navy Days, when over 21,000 members of the public came on board.

Early in September 1972, with 42 Commando, RMs embarked, the ship once again sailed for the Mediterranean and exercises off Cyprus. Before departure *Bulwark* was visited by the Commander-in-Chief, Fleet, Admiral Sir Edward Ashmore.

After the exercises, the ship sailed for Toulon, accompanied by the assault ships *Fearless* and *Intrepid.* Then she disembarked 41 Commando, RMs at Malta before returning to Devonport in mid-December.

At about this time there were suggestions that *Bulwark* might 'retire' in 1973. BEA Helicopters Ltd suggested that she could become a heliport in the Port of London. This was not to be.

Sailing from Devonport, past the discarded **Eagle.**

A full hangar. Sea Kings are seen on board for the first time.

At Toulon—the "heavies" meet. HM Ships **Fearless, Intrepid** *and* **Bulwark.**

10

Ninth Commission
1972 to 1974

The North Atlantic in winter—both crossings were memorable. It sorted out the men from the boys!

Between November 1972 and March 1974 *Bulwark* completed three deployments to the Mediterranean and two to the Caribbean/Western Atlantic area. Both crossings of the Atlantic to the Caribbean were beset by bad weather, and the hurricane force winds and mountainous seas which the ship encountered in January 1974 were possibly the most severe of the ship's career, causing extensive damage to the superstructure and cable- and quarter-decks.

Freak weather was a feature of this commission because when the ship visited Charleston in South Carolina in February 1973, a blizzard was experienced, the first in that part of the world for many years.

Bulwark had left Portsmouth in January 1973 and called in to Plymouth Sound on the 19th to embark 40 Commando. The ship headed for America and the Caribbean to take part in NATO and other exercises with the US Navy.

Bulwark became the biggest warship to have visited Charleston, South Carolina, when she sailed up the Cooper River, clearing the underside of the Cooper Bridge at mean low water by less than 8 feet.

On 9 February, 7.2 inches of snow fell, the most in Charleston in 100 years, curtailing visits to the ship. During the visit the Flag Officer Carriers and Amphibious Ships, Rear Admiral R.D. Lygo, hosted a Joint Naval Warfare Conference in *Bulwark* attended by over 40 RN and USN Officers.

After sailing from Charleston at 1130—a time selected to allow enough water under the keel to leave the jetty,

"Cooling off" in the Caribbean. The pre-wetting system is tested.

but when the tide was low enough for the ship's mast to clear the underside of the bridge—the ship re-embarked 40 Commando Group from Camp Lejeune, North Carolina, and sailed for Vieques in the Great Antilles Group, east of Puerto Rico. Here the ship and 848 Squadron trained for 'Rum Punch 73'.

After a short rest period at Roosevelt Roads, the US Naval Base on Puerto Rico, the ship sailed to start the 6 day exercise.

On completion *Bulwark* visited San Juan, the capital of Puerto Rico, before sailing home to Plymouth. On passage the ship picked up an SOS from the 18,000 tons sulphur carrier *Naess Louisiana* seeking urgent medical attention for a Portuguese seaman suffering from multiple stab wounds in his lung, inflicted during a fight over a mug. Helicopters from 848 Squadron took off in darkness to lift the man from his ship and take him to *Bulwark* for an operation.

Bad weather on passage delayed a replenishment planned for 19 March—with 'disastrous results'. The beer issue had to be limited to one can per man! The stocks had been depleted after too much had been issued for banyans at San Juan.

Five weeks were spent at Devonport, before sailing for the Mediterranean on 25 April. On passage *Bulwark* increased speed to 26 knots to look for the Russians' new Kara Class *Nicholaiair,* believed to be off the Spanish coast.

There was large scale trouble ashore in Malta when *Bulwark* arrived there to embark 41 Commando—no beer or minerals, just Marsavin.

A former Commanding Officer, Rear Admiral J.A. Templeton-Cotill, was one of the VIPs on board *Bulwark* during an amphibious exercise with *Intrepid* and 41 Commando Group, RM, off Cyprus. After the exercise *Bulwark* sailed to Athens for a 5 days visit, 2-7 June, during which Admiral Templeton-Cotill—then Flag Officer, Malta and NATO Commander South East Area

On passage to Athens. A R.A.S. with **Resurgent** *and* **Olmeda** *June 1973.*

A "full house".

Harrier trials off Cyprus.

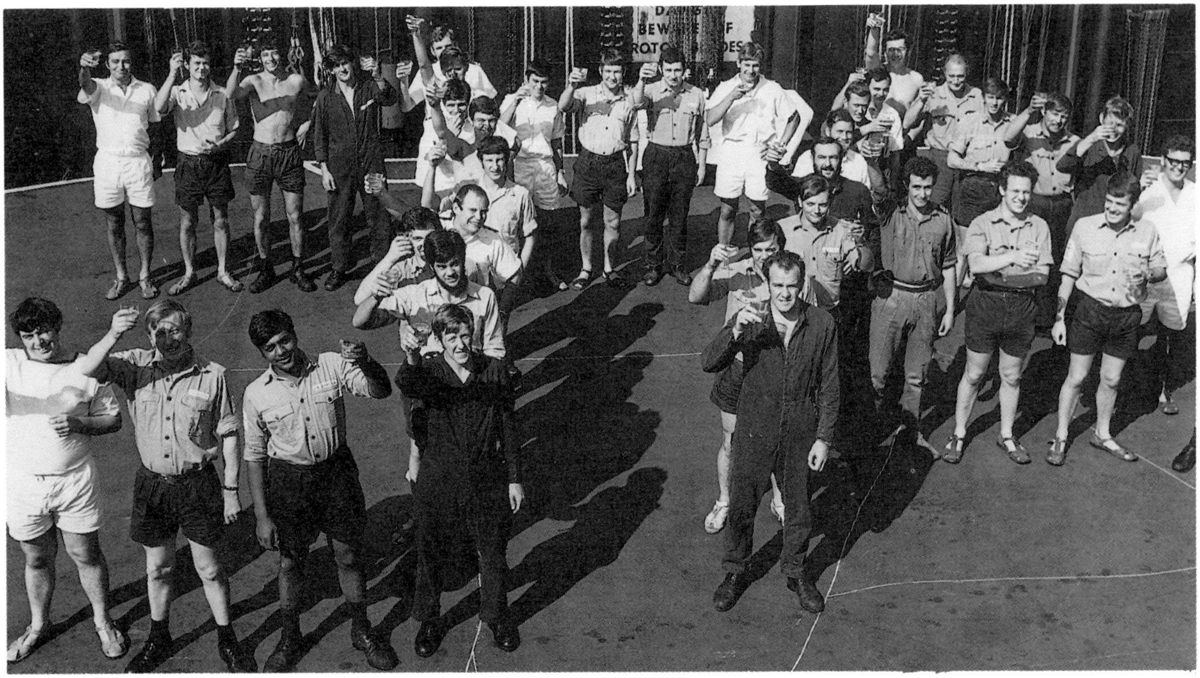

The Fleet Air Arm always like a good celebration. Here the 2,000th flying hour of **Juliet Alpha** *848 Squadron.*

Mediterranean—donned his 'NATO hat' and paid a call on the US Navy Admiral who commanded the Eastern Mediterranean Forces.

A week's self-maintenance at Gibraltar followed Exercise 'Dawn Patrol'. Then she sailed for the North Atlantic, exchanging most of 848 Squadron's Wessex helicopters for Sea Kings of 814 and 820 Squadrons off Cornwall. Just south of the Arctic Circle she took part in an Anglo-US exercise operating in the A/S role. She crossed into the Circle on 6 July.

Later that month, Defence Ministers of NATO countries, members of the NATO Council and the Military Committee, and senior members of the international staff, were the guests of Lord Carrington, Secretary of State for Defence, at a demonstration of British Maritime capability.

The joint Royal Navy-Royal Air Force display of strength and versatility, named 'Sally Forth 73', involving the largest force of HM ships and RFAs seen together for a long time, was held in the Firth of Forth.

The demonstration began with a mine hunting and fishery protection boarding exercise followed by a simulated submarine hunt by helicopters and a submarine attack by HMS *Otter* on the destroyer *Bristol.* The day ended with two Harrier aircraft taking off from *Bulwark.* While the RN awaited the outcome of the Government's deliberations on a maritime version of the Harrier, these 2 RAF 'Jump Jets' demonstrated their ability to lift off vertically—from the deck of a carrier.

From 26 July *Bulwark* was at Devonport, but on 4 September sailed again for warmer climate, this time with 42 Commando. During the latter part of the month, *Bulwark* took part in 'Deep Furrow 73', an annual exercise scheduled by the Commander-in-Chief Allied Forces, Southern Europe, in the eastern Mediterranean. In addition to *Bulwark*, other UK Naval units taking part included HM Ships *Intrepid* (Commodore D.T. Smith, Commodore Amphibious Warfare) and *Devonshire*; and the logistic landing ships RFA *Sir Lancelot* and *Sir Galahad.*

'Deep Furrow 73' was designed to test the capabilities of the forces on the southern flank of the Alliance and involved some 54 ships and supporting forces, and over 50,000 soldiers, sailors and airmen from the United States of America, United Kingdom, Greece, Turkey and Italy.

The naval escort forces initially assembled off south west Turkey to carry out anti-submarine, gunnery and carrier borne aircraft exercises, while the combined amphibious assault forces under the overall command of Vice-Admiral C.E. Bell, USN in the USS *Mount Whitney,* proceeded north to Saros Bay in north west Turkey to carry out a landing in support of visits of the Turkish First Army ashore in action against the 'enemy' Orange forces. Allied aircraft carried out strikes inland before the assault and provided air-cover over the beachhead, as over 2,000 Royal Marine Commandos and United States Marines went ashore. More than 400 Royal Marine Commandos and US Marines were also landed by helicopter and a massive airborne drop involving Turkish, British and American paratroopers was made in support of the operation which forced the Orange forces to withdraw. Throughout the assault the supporting naval forces provided naval air cover and protection against any Orange force surface or submarine threat.

After the exercise, the ships returned to Istanbul for the inevitable 'wash-up'.

Bulwark afterwards returned to Devonport to drop 42 Commando and then proceeded to Rotterdam to embark the 1st ACG (Amphibious Combat Group) of the Royal Netherlands Marine Corps. On departure from Rotterdam the weather became so rough that the river pilot had to remain on board till the ship reached Portland. After a logistic exercise off Portland, the ship returned to Rotterdam to disembark the 'cloggies'. Before turning for home, the ship made a 7 day visit to Travemunde where, despite the ice and snow, the welcome was very warm.

On her return to Devonport on 13 December she was able to enter harbour before first light, using new navigational aids fitted ashore, the first major war vessel to use the system.

En route to her 1974 Caribbean deployment, with a force of Royal Netherlands Marine Corps who had been embarked off Valkenburg, *Bulwark* ran into heavy weather about 300 miles north east of the Azores. Winds gusting over 70 knots generated seas 30-40 feet high and repeatedly buried the forward area of the flight deck, damaging catwalks.

On 16 January the sea state code reached 9—phenomenal—and the ship was pooped when she drove deep into a large trough. The bridge met green seas and the quarterdeck was under water. The lifebuoy sentries were found paddling their raft in the Commander's flat. That evening the ship rolled over 23°, adding to the damage inside the ship. The next day was calmer and the ship fuelled and 'vertrepped' beer from RFA *Tidepool*, but it was not till the 21st that the Captain's carpet could be taken to the flight deck to dry out.

The jumbo crane was used to put cutting equipment over the side, to remove twisted metal, before the ship's arrival in the Caribbean.

The Queen Emma floating bridge swung aside to allow *Bulwark* to enter the harbour of Willemstad, Curacao, in the Dutch Antilles. There the Dutch Marines went ashore for exercises and tropical acclimatisation before joint UK-Netherlands-US amphibious exercises off Puerto Rico.

Bulwark also called at Cartagena, Colombia's main naval port, being the first ship of her size to do so. A guard and band from the ship marched through the city streets to Bolivar Square, where her Commanding Officer (Captain D.W. Bazalgette) laid a wreath at the Statue of Liberator, Simon Bolivar.

When the weather moderated fuel—and beer—were embarked from RFA **Tidepool.**

Calmer waters at last—entering Curacao.

Time to relax at last. On the beach at Cartagena.

As ever—children were entertained on board at the traditional pirates' party . . .

. . . and 2A Mess help celebrate a messmate's birthday.

Fun in the sun. **Bulwark** *at Aruba, February 1974.*

In mid-February 1974 fourteen RN and Royal Fleet Auxiliary ships, with some 5,000 sailors on board, anchored off Virgin Gorda, one of the British Virgin Islands in the Caribbean. The ships were there for a series of routine NATO, bi-national and national training exercises, code-named 'Caribtrain'. The gathering provided the opportunity for competitions and recreation in the sun and for the First Sea Lord, Admiral Sir Michael Pollock, GCB, MVO, DSC, to say farewell to the Fleet before he retired at the end of the month.

A busy programme of ships' visits was carried out by the First Sea Lord and the Commander-in-Chief Fleet, Admiral Sir Terence Lewin, KCB, MVO, DSC. *Bulwark*'s flight deck was the venue for the sports competitions for which Lady Pollock presented the prizes. In the evening there was a buffet on the quarterdeck for 100 officers from the assembled ships whilst 'horse racing' was conducted on the flight deck.

The ships left the anchorage on the morning of Friday 15 February with the First Sea Lord flying his flag in *Bulwark,* and as the assembled fleet, led by the submarine *Narwhal,* steamed past the Commando Carrier, their ships' companies cheered ship. Once clear ahead they divided into two divisions and turned to pass either side of *Bulwark*, co-ordinated with a fly-past by Sea King, Wessex and Wasp helicopters.

Before departing the First Sea Lord made the following signal—"Thank you for giving me the most marvellous send-off and a final view of the Navy which I shall never forget".

Bulwark then left the other ships to continue her exercises with the Dutch Marines. After a visit to Oranjestad in Aruba she sailed back across the Atlantic. On 5 March there was a full day's 'backras' with RFA *Regent* to prepare the ship for refit. Two days later, as the ship was heading up Channel in fog to disembark the Dutch Marines, a fire broke out in 'Q' diesel generator compartment. Although it was a major fire, the marines were able to disembark that day, and the ship returned to Devonport the next day, 8 March, for a refit aimed to improve the living accommodation.

During the refit the Chinese LEP cooks and stewards all took the opportunity to enjoy conducted tours over Dartmoor and sample Devonshire fare.

11

Tenth Commission 1974 to 1976

Not a pretty sight . . . in dock at Devonport.

Although there was no formal ceremony, the tenth—and what was expected to be the last—commission of *Bulwark* began when her new Commanding Officer, Captain J.R.C. Johnston, assumed command on 18 June 1974. The ship was still undergoing refit which, for a number of reasons, was not completed until 7 February, 1975. Even on the day she was due to sail on her post-refit sea trials, a joint in the superheated steam system 'blew out' and delayed her departure for 24 hours.

This was to be her farewell to Devonport because it had been announced at the end of January 1975 that in future her 'home port' was to be Portsmouth.

After sea trials, she sailed on 10 March for a work-up off Portland before leaving home waters on the 14th for the Mediterranean. Shortly after entering the Mediterranean,

Farewell to Devonport.

Work Up . . .

Two Sea Kings of the Royal Australian Navy carry out deck landing practice.

The Ship's "main armament" fire a broadside!

Seamen and aviators share the flight deck during work-up evolutions.

Pilots of tomorrow? A Sea Cadet Unit pose in front of the latest technology to arrive onboard.

The down draught from a visiting Harrier gives the ship's side a good wash.

the helicopters of 848 Squadron were called upon to assist the frigate *Ashanti* who was towing the bow section of the tanker *July Star* back to Gibraltar. When *Bulwark* arrived on the scene, one of her Wessex aircraft assisted the tug *Robust* to take over the tow; then all the ships made their way to Gibraltar. From there *Bulwark* made her way to Malta for a 3 weeks maintenance period. Towards the end of the period in Malta, the ship's divers discovered a number of cracks in the starboard propeller and plans were put in hand for docking the ship in Malta. However, it was decided not to change the propeller until after the

Gibraltar for a quick propeller change.

exercise which was planned to shortly take place in Sardinia.

On 18 April *Bulwark* sailed to rendezvous with units of the US Sixth Fleet and the Italian Navy off Cape Teulada in south west Sardinia. Weather conditions restricted activities during the exercise and the commando landings had to be carried out by helicopters only, and not by landing craft.

During the exercise it was learned that the propeller change would not take place in Malta after all, but at Gibraltar. The ship arrived there on Friday 25 April, went into dock the same evening, had the propeller changed on Monday and sailed again on Wednesday morning.

After a quick passage to Malta, where the Commandos were disembarked, preparations were made for a visit to the French Riviera. As a RM Band was not available, the Pipes and Drums from the Band of the 1st Battalion Irish Guards were flown from Germany—courtesy of the RAF—to Sardinia and then transferred to the ship by 848 Squadron helicopters. An old naval custom was also revived when a Maltese dghajsa and crew were lifted onto the flight deck to join *Bulwark* for the trip.

The Pongos perform . . . The pipes and drums of the 1st Battalion Irish Guards on arrival at Villefranche.

The 30th anniversary of VE Day was suitably remembered.

May 1975. **Bulwark** *and RFA* **Tidepool** *become close friends.*

The official visit to Villefranche was 'cunningly contrived' to coincide with the Monte Carlo Grand Prix and the Cannes Film Festival. *Bulwark* arrived on 7 May 1975 and a few days later took part in celebrations for the 30th anniversary of VE Day. Some of the ship's company and Royal Marines, and the Pipes and Drums of the Irish Guards took part in the march past.

On passage back to Malta, medical assistance was provided to the merchant ship *Northern Joy*. *Bulwark*'s SMO was winched down to the aid of a Greek seaman who was later transferred to *Bulwark*'s sick bay, and, on the ship's arrival in Malta, to a hospital on shore.

After a week in Malta, 350 men of 41 Commando were embarked for an Amphibious Training Exercise 'Double Bass' which was to take place off north Corsica. The Commando's opposition was provided by a company of French Foreign Legion. When they visited *Bulwark* for familiarisation, they were received with awe by the ship's company. Once again the weather was to play an important part and the full extent of *Bulwark*'s capabilities could not be shown.

A brief visit followed to the small port of La Maddalena—only a small place but one of historic interest in that Nelson used it as an anchorage during his blockade of the French Fleet at Toulon in 1803.

Then it was back to Malta for *Bulwark* to embark the whole of 41 Commando—750 men—and a company of the 1st ACG of the Royal Netherlands Marine Corps for the biggest exercise in the Mediterranean during the year—'Dawn Patrol 75'. This was designed to improve the deterrent and readiness posture of the southern region and provide for advanced training in naval air and amphibious operations. It involved about 50 ships from the RN, USN, Italy and Turkey. They met in Taranto Bay from where they sailed on 19 June to commence the transit phase of the exercise, making an opposed passage round southern Sicily, before a rendezvous to the north west with the slower RFAs who had passed through the Straits of Messina. The main assault off Cape Teulada took place at dawn on 23 June.

41 Commando and the Dutch Marines were landed in Malta before the ship proceeded to Gibraltar. She was ordered to return to Malta for one of the all too many crises occurring in the eastern Mediterranean. However, a few hours later another signal was received instructing *Bulwark* to return to Portsmouth. 848 Squadron was disembarked as the ship headed up Channel, and she anchored at Outer Spit Buoy on Thursday 3 July. The next day was spent at sea—for Families Day—before berthing alongside Pitch House Jetty in Portsmouth.

Back to Portsmouth—to be met by wives, sweethearts . . . and other young ladies . . .

. . . and the 1975 Navy Days crowds.

Just like the good old days. The Royal Navy returns to Grand Harbour in force.

After summer leave—and 33,000 visitors to the ship during Navy Days—*Bulwark* left Portsmouth on 1 September. With 848 Squadron and 40 Commando, RM on board (the latter embarked at Plymouth) she headed again for a fast passage to the Mediterranean. Only a few hours were spent in Gibraltar Bay and a brief visit to Malta before leaving there on Monday 15 September for the NATO Exercise 'Deep Express 75'. Next day *Bulwark* anchored off Antalya in southern Turkey where the remainder of the Task Force—UK, US, Italian and Turkish ships—had assembled for a pre-exercise conference. The pattern of 'Deep Express' was similar to 'Dawn Patrol' in that the Task Force made an opposed transit through the island-studded Aegean Sea, and then made an early morning assault on beaches in Turkish Thrace. Throughout the exercise, the ships were under close surveillance by Russian ships. After being 'bombed, strafed and torpedoed' with monotonous regularity, the assault force was landed on the north side of Saros Bay on Monday 22 September.

On completion of the exercise *Bulwark,* in company with *Hermes* and other HM ships, sailed to Istanbul for a five day visit.

Istanbul is truly one of the great cities of the world and one of the very few that could absorb as many men as were landed from the British and Italian ships anchored in an impressive line just opposite the Dolmabache Palace. Whether one's interests were in ancient history, shopping in the teeming bazaars, sampling unusual meals or seeing the Oriental night life—complete with genuine belly dancers—there was undoubtedly something for everybody. It also soon became apparent why foreign visitors were recommended not to drive in Istanbul. The existence of a highway code in Turkey was considered debatable, and if it did exist perhaps it had never been promulgated to the citizens of Istanbul. The term 'aggressive driver' took on a new significance here.

All the ships left Istanbul by 5 October to avoid a clash with the end of the fast of Ramadan. *Bulwark*, with *Hermes*, proceeded up the Bosporus and turned almost within sight of the Black Sea, much to the consternation of the local fishermen who found themselves encircled by two huge carriers.

After an uneventful passage through the Dardanelles and Aegean Sea, *Bulwark* reached Malta on 7 October. A few days were spent there before again heading home, arriving in Portsmouth on 21 October.

Berthed alongside Pitch House Jetty, the thought uppermost in most minds was FOCAS's Administrative Inspection. At times, it was almost impossible to move about the ship due to passageways and flats being closed for painting and polishing. Coincident with Admiral Eberle's personal inspection of *Bulwark*, was the ship's 21st birthday—the anniversary of the ship's first commissioning ceremony in October 1954. CPO Cook Cheng Wing and CPO Cook Ken Reid produced a magnificent iced cake, to celebrate this coming-of-age, with the names of all the aircraft, all the Squadrons, all the Commando Groups and many of the places visited by *Bulwark* on the cake.

Bulwark *comes of age . . . and a magnificent cake is baked to mark the occasion.*

In November more than 2,000 American, Netherlands and Royal Marines took part in the Exercise 'Triple Jubilee' which was designed to practise the capability of Joint Amphibious Forces.

It began on 31 October with the US Amphibious Task Force arriving at Plymouth where for a week the US 36 Marine Amphibious Unit carried out individual and sub-unit training on Dartmoor, and with Royal Marine units from Plymouth and the Exmouth area. On 7 November the Force re-embarked and carried out an assault landing

The rumours start in the UK press. Is this really the last commission?

at Lulworth in Dorset. The next day they moved into Salisbury Plain for a series of combined operations. After stiff opposition from the 'enemy' forces, provided by elements of 40 Commando, RM, it was decided to call up reinforcements and the First Amphibious Combat Group, Royal Netherlands Marine Corps from *Bulwark*, lying off Portland, were flown in by 12 Wessex Mk 5 helicopters of 845 NAS at first light on 11 November.

Christmas was spent in Portsmouth but early in January 1976 *Bulwark* was off to sea again. 848 Squadron flew in from RNAS *Yeovilton* to join the ship on the 6th, and marines from the Netherlands Amphibious Combat Group were embarked, together with the Band of the 3rd Battalion Royal Green Jackets. At Plymouth, 750 men of 40 Commando RM were picked up before, in company with RFAs *Tidereach, Retainer, Olwen* and *Regent, Bulwark* sailed on the 9th for the Caribbean.

On passage across the Atlantic, a helicopter of 848 Squadron was forced to ditch. Fortunately the crew were rescued unhurt before the helicopter sank.

Now how did that happen, Chief?

Nothing to do with us!

Nassau, February 1976. Cruise ships all . . . for one ship the routine was somewhat different once at sea.

Success at last. The ship's anglers are rewarded.

During the next two months *Bulwark* carried out a bi-national exercise with Dutch Forces off Curacao; joined the US Marine Corps in an exercise off Puerto Rico; and carried out an exercise in the Bahamas with the Marine Wing of the Royal Bahamian Police Force.

The highlight of the deployment was a visit to Fort Lauderdale in Florida, where 848 Squadron performed a fly-past over the sunsoaked beaches. The ship also visited the island of Aruba in the Dutch Antilles, where the Lieutenant Governor of the island lunched on board.

As part of the cuts in defence spending, it had been decided that *Bulwark*'s disposal date was to be brought forward two years to March 1976, but the following week it was announced that she was to be kept in reserve for up to 5 years.

On the way home to Portsmouth, *Bulwark* called at Gibraltar; flew off 848 Squadron while the ship was off Cornwall; made a brief stop at Plymouth to disembark 40 Commando; and anchored off Portland Naval Base to de-ammunition.

A ceremonial entry was made into Portsmouth with a paying-off pennant flying; the ship's Band and that of the 3rd Battalion Royal Green Jackets playing; and the helicopters of 848 Squadron flying overhead. One of the ship's officers played a lament on the bagpipes. As a press release at the time said "after 21 years of memorable and, at times, momentous service in the Fleet, it can truthfully be said that this Old Lady has lifted her skirts and enjoyed her final fling."

On 9 April over 500 guests, including six of her previous Commanding Officers, attended a Farewell Ball in the ship's hangar. Within a few days the ship's company had been reduced to little more than 200 whose job was to keep her prepared for further service if required.

12

Preservation by Operation
1976 to 1978

In May 1976 *Bulwark* was put into a type of Reserve known as 'Preservation by Operation'. To maintain the ship in this state of hibernation required a ship's company of about 200 officers and men, mostly technical personnel, under the command of Commander A.S. Carpenter, RN.

In 1976 the Peruvian Navy showed an interest in acquiring *Bulwark*, with a view to operating her with Harrier aircraft. A delegation of senior Peruvian officers inspected the ship as she lay at Portsmouth. However, the venture came to nothing because the Admiralty Board were unable to give a firm date for her sale.

Late in 1977 rumours became rife that *Bulwark* would be re-commissioned, spurred on by the news that *Ark Royal* was to pay off in 1978 and that the new class of carrier would not be ready for service until the early 1980s. Many months of speculation followed, until Mr Fred Mulley, Defence Secretary, announced in January 1978 that she was to be brought forward into a fully operational state. Her primary role was to be anti-submarine work, operating Wessex Mk 5 and Sea King helicopters, but she would still retain her ability to carry Commandos.

On 20 March 1978 she was taken in hand at Portsmouth for a 10 months programme to prepare her for service. Many problems were encountered by the technicians—not only because her equipment had been out of use for some time, but also because of its age. Much of the equipment had been fitted during the original building period thirty years before. No longer were the items commonplace, and many of the electrical units were the old direct current type instead of the modern alternating current.

13

The Final, Final Commission 1979 to 1981

H.R.H. The Prince of Wales cuts the final commissioning cake.

Despite the difficulties, *Bulwark* was reactivated—her task, which was expected to be for five years, was to fill, together with her bigger sister *Hermes*, the period until the new *Invincible* class could join the Fleet.

The ship was re-commissioned on 23 February 1979 under the command of Captain G.M.K. Brewer. The principal guest was HRH The Prince of Wales, who, assisted by the youngest member of the ship's company, JMEM Michael Dwyer, cut the commissioning cake. Amongst the guests was Admiral of the Fleet The Earl Mountbatten of Burma, who remarked that he had never seen the ship looking more 'spick and span'. Prince Charles must have been impressed by the cake, made by Chief PO Cook Dave Avery, because when the Prince was married, Chief Cook Avery made his wedding cake and, later, made the christening cake for Prince William.

Four days after recommissioning, *Bulwark* left Portsmouth for trials, shakedown and work-up, after a spell of almost three years in preservation alongside. 'Flying Stations' was sounded as she passed the harbour entrance and the Sea Kings of 826 Squadron were launched for the first time at sea, accompanied by Wessex aircraft of 846 Squadron as *Bulwark* passed the Outer Spit Buoy. At anchor that night, a signal was made to inform the Principal Guest at the recommissioning, that after a successful day of full flying operations, the

Bulwark arrives at Rosyth for 1979 Navy Days, with Sea King helicopters lining her flight deck.

"mainbrace had been spliced" as commanded. The remainder of the trials went well too, including a full power trial when over 28 knots were achieved. The ship was declared operational two days earlier than originally scheduled a year before. Her first exercise in the new anti-submarine role proved her capability and Whitsun was spent at Rosyth as the main attraction for Navy Days.

Shortly afterwards *Bulwark* visited Hamburg which enabled the ship's company to exchange duties with members of the ship's affiliated regiment, the 4th/7th Royal Dragoon Guards, who were based only a few hours drive away. HMS *Bulwark* cap tallies were to be seen driving Chieftain tanks on the ranges, while men of the 4/7 DG drove flight deck tractors; carried out armouring duties and even joined the galley crew.

On 31 May, the ship and her Squadrons completed a successful operational readiness inspection. Early in June, 350 members of 42 Commando were embarked to enable the ship to practice her alternate role as a Commando Ship.

Whilst on passage to Copenhagen for a very welcome break between the first and second Commando work-ups, the ship diverted to help in rescue operations when a French trawler sank after colliding with a merchant ship. Sadly, the crew were already beyond assistance but one body was recovered from the sea.

On 28 June, 45 Commando, RM embarked from their base at Arbroath and the ship proceeded to Den Helder to pick up the 1st Amphibious Combat Group of the Royal Netherlands Marine Corps. Then, with aircraft of 820 Squadron joining the air group, *Bulwark* headed back into the Baltic for a major Amphibious Exercise 'Whiskey Venture', centred on one of the Danish islands.

July and August were spent in Portsmouth carrying out routine maintenance, but on 19 September the ship set off for an autumn deployment in the Mediterranean. It began with a three week spell at Gibraltar to catch up on the backlog of planned maintenance and, of course, allow for some recreation. The band of the 4/7 DG had been embarked for passage to Cyprus, to join other members of the Regiment already there. As the ship headed into Gibraltar Bay, the band was paraded for the salute. It was then found that the music for the salute had been mislaid but the situation was recovered when several keen Gilbert and Sullivan fans among the ship's company, mustered to sing the tune while the bandmaster wrote it down. It was said that the result was indistinguishable from the original—at a distance of two miles!!

On the morning of 18 October, with the ship anchored off Akrotiri in Cyprus, men of the Army Air Corps and the 1st Battalion The Queen's Lancashire Regiment joined the ship, to be taken around the coast to take part in an exercise code-named 'Sea Rose'. With members of the RAF Regiment acting as the 'enemy', the men of the QLR were landed by helicopter and LCA.

Once the exercise had been completed the ship's company were able to relax for two days, before the ship sailed for a five day visit to Naples. There the Chaplain's

The first deck landing onboard of a Sea King MK IV Helicopter—4th January 1980.

(Roger Bennett) famous magical mystery coach tours—known as 'Bish Vista Tours'—really came into their own and later they became a major feature of other visits.

After a brief stop at Barcelona, *Bulwark* returned to Portsmouth, where she arrived on 22 November. Two Sea Kings were given clearance to fly back from the Mediterranean, overland as a training exercise.

On 4 January 1980, having taken over the mantle of the front line carrier for the Royal Navy, and with 814, 826 and elements of 846 Squadrons embarked, *Bulwark* sailed for her North American deployment.

Mr David Shepherd, the famous elephant and steam locomotive artist, was on board, for, having been invited to paint *Ark Royal* for the Fleet Air Arm Museum, he

Another ditching, but the crew were saved.

needed some good photographs of a carrier under way at speed. *Bulwark* obliged in the South Western Approaches before heading out into the Atlantic.

During intensive flying operations, the ship completed her 4,000th deck landing since recommissioning and although one Sea King was lost after ditching, her aircrew were all recovered safely to maintain a fine flight safety record.

The ship spent a five week Contract Assisted Maintenance Period at the US Naval Station, Mayport in Florida. While the ship was in the hands of the contractors, the three Squadrons of helicopters were disembarked to the Naval Air Station at Jacksonville. Marvellous facilities were provided for the many families who flew to the US to join their menfolk, and many of the unaccompanied members of the ship's company took the opportunity to explore other parts of the USA.

During the visit to Mayport, Captain P.J. Symons relieved Captain Brewer, and the change of command ceremony attracted much attention from the Americans.

Bulwark sailed from Mayport in sunny weather on 22 February to take part in Exercise 'Safe Pass', scheduled to last from 24 February to 7 March. The day after the exercise ended she entered the port of Philadelphia for a nine day stay. During the early hours of the 15 March, a serious hydrogen fire started in 'A' boiler room and within a short time A1 boiler was completely burnt out. By midday though, the ship's fire party, assisted by the Philadelphia Fire Department, had managed to put out all the fires. Then a big clean-up began; the inside of 'A' boiler room was completely blackened, a thick layer of soot covering everything.

The ship sailed using only three boilers and then, in mid-Atlantic, 'B' boiler had also to be shut down to enable repairs to be carried out to a large steam leak. The ship drifted about 29 miles whilst repairs were completed. She finally arrived at Portsmouth on 2 April for Easter leave.

Three weeks later *Bulwark* sailed to Arbroath to embark 45 Commando. Then followed anti-submarine exercises in the North Sea before she headed for the sunnier climes of the Mediterranean and Exercise 'Dawn Patrol' on the coast of Sardinia.

A courtesy visit to Toulon preceded a three week docking period in the massive Prince of Wales Dock at Gibraltar. After undocking, the ship again returned to Portsmouth for a combined leave and Dockyard Assisted Maintenance Period. There she remained until the middle of August, when anti-submarine exercises were carried out in the Channel before she called into Plymouth to embark 40 Commando. While they were carrying out exercises in Norway, *Bulwark* took part in an anti-submarine exercise in the North Atlantic, 'Teamwork 80'—held in typical North Atlantic weather; rain, mist and heavy seas.

The discomforts of the grey Atlantic were soon forgotten when the ship arrived at Hamburg to enjoy the local hospitality, from 27 September to 3 October 1980.

Between exercises the ship spent a weekend, 17-20 October at the French port of Brest. After anti-submarine exercises with the Dutch Navy in the Western Approaches and a brief visit to Gibraltar, *Bulwark* returned to Portsmouth.

On 9 November another fire occurred on board—this time in the forward hangar and several messdecks. Whilst the ship's company were clearing up, they learned that *Bulwark* was to be taken out of service and placed on the Disposal List at the end of March—six months earlier than expected. The real disappointment was that the last deployment to North America was cancelled and her only visits abroad would be to Gibraltar and Lisbon. These were followed by 'Cold Winter'—an exercise off the Norwegian coast with the men of 42 Commando embarked—the same Commando which had been on board when *Bulwark* was first commissioned as a Commando ship in 1960. They were landed at Plymouth before *Bulwark*—a victim of defence cuts—docked at Portsmouth for the last time on 27 March 1981.

Home at last . . . 27th March, 1981. **Bulwark** *arrives at Portsmouth.*

Procedure Alpha for the very last time.

Jack's humour—as **Bulwark** *arrived alongside, a little advertising was allowed.*

"Heavyweights' Corner" in Portsmouth Naval Base. **Bulwark** *lies between HM Ships* **Intrepid** *and* **Fearless**. *Soon the Falklands crisis would break . . . but it was eventually decided that* **Bulwark** *would not be involved.*

14

Final Days

Deserted . . . a once proud ship **Bulwark** *lay at this buoy for 2 years awaiting a decision on her fate.*

For twelve months *Bulwark* lay in a corner of No. 3 Basin in Portsmouth Dockyard, being de-fuelled and de-stored. Frequent visits were made to her by members of the Portsmouth Fleet Maintenance Group who removed items for use in other ships.

Thereafter *Bulwark* became a familiar sight, moored in the upper reaches of Portsmouth harbour—waiting hopefully for yet another call back to active service. This time the call was not to come.

She was nearly reprieved during the 1982 Falklands War, when tentative plans emerged to tow her to the South Atlantic as a floating barrack block and heliport for the British troops. But that scheme foundered and she remained a sad sight by the mudflats.

From the Captain's Cabin, the photographer catches HM Ships **Tiger** *and* **Kent**—*also awaiting their final voyage to the scrapyard.*

Scenes of Desolation . . .

The Forecastle.

The Captain's Cabin . . . many VIP's were entertained here. Now it is gutted ready for the woodwork to be burned—and the steel cut into indistinguishable sections.

The Wardroom . . . the same sad tale . . . if only these bulkheads could talk they could tell a story or two . . .

The once proud and gleaming quarterdeck is rigged for the tow—the brass work no longer gets its daily polish as rusty cables litter the once gleaming deck.

The Upper Bridge. No views of Singapore, Hong Kong, Fremantle or Cape Town now—just the mud flats of Fareham Creek.

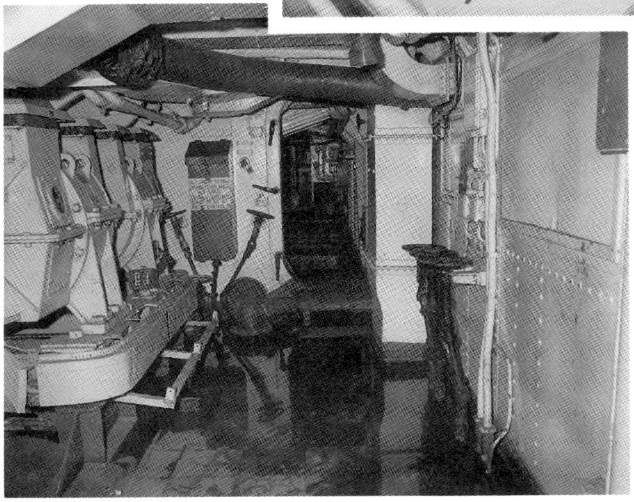

The Burma Road . . . the main passage way through the ship once thronging with sailors and marines as they attempted to squeeze past the human traffic going in the opposite direction. Always a bottleneck here . . . the flashlight from the camera reveals a small flood in the passage way.

On Tuesday 10 April 1984 three former Commanding Officers, led by Rear Admiral Percy Gick, joined a handful of HMS *Bulwark* Association Members gathered at Portsmouth to say farewell to 'their ship'. Just to add to the poignancy of the parting, they had to witness *Bulwark*'s last exit, from the flight deck of USS Saipan, an assault ship in the military role which *Bulwark* had pioneered more than 20 years before. *Bulwark* was on her way to the ship-breaking yard of Messrs Shipbreakers (Queenborough) Ltd.

The harbour tugs *Adept, Bustler, Foxhound* and *Dalmatian* eased her from her moorings, and the ocean-going tug, RMAS *Robust*, led her out of harbour. Once past the Outer Spit buoy the last named took up position on *Bulwark's* starboard bow, with RMAS *Rollicker* on the port bow, in readiness for the long haul to Cairnryan, near Stranraer.

Heading for the Graveyard

Off the Devon coast . . . the Tugs **Robust, Rollicker** *and* **Capable** *lead the "Old Lady" to Scotland.*

Her half sister, **Hermes,** *passes to pay her last respects.*

Arrival at Cairnryan. The pile of scrap in the foregound (from HMS **Blake***) indicates the future of the mighty* **Bulwark.**

In the early hours of Wednesday morning, *Rollicker* suffered a complete engine failure and had to slip her tow. The towing master signalled ahead to Plymouth to inform the authorities that *Rollicker* could only be considered "unreliable" and another tug should be allocated or *Bulwark* would have to be taken into Devonport. By 0600 the crew of the tug *Capable* had been roused from their homes and their vessel was on its way to join *Robust* and *Bulwark*.

By the time *Capable* had joined, *Rollicker*'s problems had been solved and she was able once again to pass the tow to *Bulwark*. However, *Capable* remained on station and took up a position about a mile ahead to guard against shipping that might come too close.

Such good headway was made that it first seemed possible *Bulwark* would arrive at Cairnryan on Friday the 13th—24 hours ahead of schedule. However, when she was less than 20 miles from Cairnryan, it was decided that she could not make the Friday morning tide, so a reduction in speed was ordered. This resulted in *Bulwark* starting a "controlled" drift off the Irish coast and, although the tugs were steaming ahead, they were actually going astern over the ground. By midnight they had covered 114 miles in 24 hours—most of it backwards!

The plan to arrive on the Saturday was cancelled by the pilot on shore, once he saw the weather forecast for that day. Sunday was also spent at sea, with the 'Old Lady' tugging at her wires as if she knew she was heading for the cutting torches; she seemed to be delighted with her few days extra stay of execution. With the sea and tide to help her, she spent the afternoon flapping at the end of her wires—sometimes at almost 90° from the tow, as the wind caught her.

High tide at Cairnryan on Monday 17th April 1984, was about 1230, at which time *Bulwark* was at the entrance to the Loch. A civilian tug from the Clyde, *Flying Phantom*, replaced *Robust* on the starboard bow; *Robust* went aft to take a line from *Bulwark*'s quarter, to swing the carrier through 180° ready for her arrival alongside. She was finally berthed between what remained of the frigate *Mohawk* and cruiser *Blake* in the yard where the aircraft carriers *Ark Royal* and *Centaur* had been broken up before her.

Bulwark had arrived for her final conversion—into a mountain of scrap metal.

By October, 1984, the scrap man has really got his teeth into "the job".

Appendix One

COMMANDING OFFICERS

	Date of Appointment (As shown in Navy lists)
Captain J.M. Villiers, OBE	1 October, 1954
Captain P.D. Gick, OBE, DSC*	14 January, 1957

In 1958-59 after Captain Gick left the ship, the Senior Officer on board was Commander R.H.H. Brunner, DSC*.

Captain R.D. Franks, DSO, OBE, DSC	1 August, 1959
Captain J.S. Dalglish, CVO	6 August, 1961
Commander W.H. Hoyle	18 February, 1963
Captain P.J. Morgan, DSC	16 September, 1963
Captain D.B. Law, MBE, DSC	29 December, 1964
Captain A.M. Power, MBE	27 December, 1966
Captain J.A. Templeton-Cothill	1 May, 1968
Captain B.H. Notley	3 November, 1969
Captain L.R. Bell-Davies	15 December, 1971
Captain D.W. Bazalgette	13 December, 1972
Captain J.C.R. Johnston	18 June, 1974
Commander A.S. Carpenter, MBE	1 October, 1976
Captain G.M.K. Brewer	3 July, 1978
Captain P.J. Symons	18 February, 1980

Appendix Two

THE BRITISH INSURANCE ASSOCIATION ADOPTER AND BENEFACTOR

In April 1955, the British Insurance Association—who for many years had been associated with the previous trials and training carrier *Illustrious*—adopted HMS *Bulwark*.

Never has an organisation taken such a great personal interest in, or generally looked after a ship so well, as the way the BIA took care of *Bulwark*.

Within the first few months the BIA donated musical instruments, which inspired the formation of a Volunteer Band, and a handsome silver trophy expressly for "inter-part" competition on board. The latter was competed for on an aggregate points basis in a competition involving a number of different sports.

Numerous other presents, including a fine supply of library books, were received by the ship. The BIA was always pleased to be able to contribute to the ship's welfare fund for purposes which were not covered by normal Naval funds.

Visits to London were organised to enable members of the ship's company to spend a day visiting interesting and important places in the city, such as the Guildhall, Mansion House and Stock Exchange.

Social and, in particular, sporting activities were always well to the fore in the programme of events, not only in England, but elsewhere in the world, where British Insurance was represented.

In 1976, the BIA was presented with the Ship's Bell, on loan until the final destiny of *Bulwark* became clear. Unfortunately for BIA, the bell had to be returned rather sooner than expected. With the ship having finally gone to the shipbreakers, the bell was bought by the HMS Bulwark Association from the Ministry of Defence for £500, with the intention of presenting it to the Fleet Air Arm Museum at Yeovilton.

The BIA still maintains its connection with the Royal Navy, having become officially associated with HMS *Invincible*.

Appendix Three

The First Bulwark

1778 to 1780

A 74 guns third rate to be named *Bulwark* was ordered on 11 June 1778 to be built at Portsmouth Dockyard. Designed to be 1,620 tons (builder's measurement), 169ft long and 47ft beam, she was registered on the List of the Navy by an Admiralty Order dated 23 July 1778. Work commenced in 1780 but little was done before the Order was cancelled.

The Second Bulwark

1807 to 1826

Another 74 guns third rate, designed by Sir W Rule, was laid down in Portsmouth Dockyard as *Scipio*. Her principal dimensions were 1,940 tons (builder's measurement); length along gun deck 181ft 10ins; 49ft 3ins beam; with a crew of 590 men.

In 1806 she was renamed *Bulwark* and was launched on Thursday 23 April 1807 after being christened by Admiral Montagu.

Commanded by Captain Farmery Predam Epworth, *Bulwark* saw action in the American War of 1814-15. As one of a fleet of 8 men-of-war and 10 transports, commanded by Rear Admiral Edward Griffith in *Dragon*, she carried out an expedition up the Penobscot River. After slight resistance, the town of Castine was captured. A party from *Bulwark*, supported by men from other ships, proceeded further up river where they forced the Americans to burn their own 36 guns frigate *Adams*. They also drove the Americans out of Hampden.

On 23 October 1814 *Bulwark* captured the American 10 guns privateer *Harlequin* and on 22 July 1815 a similar ship *Tomahawk*.

Bulwark was finally broken up at Portsmouth Dockyard in September, 1826.

The Third Bulwark

1859 to 1873

This ship was to be an 81 guns second rate designed for a screw. Of 3,716 tons (builder's measurement); 252ft long and 58ft beam, she was laid down in Chatham Dockyard on 8 March 1859.

With work on the ship well advanced, an Order was received on 7 March 1861 to cease work. For 12 years she remained high and dry, before a further Admiralty Order, dated 22 August 1872, ordained that she should be 'taken to pieces'—a job that was completed by mid-March 1873.

The Fourth Bulwark

1860 to 1921

This ship was launched on 7 March, 1860 at Pembroke as the *Howe*. She was a screw wooden battleship of 121 guns, 4,245 tons and engines of 1,000 nominal horsepower. Her length was 275ft and beam 61ft. Her only sea-going cruise was that made on her steam trials in 1861, on the conclusion of which she returned to her moorings. For about 25 years she lay in the upper part of the Hamoaze at Devonport before being brought down river to perform the duties of a training ship.

On 3 December 1885, the *Howe* was renamed *Bulwark* but was renamed *Impregnable* on 27 September 1886.

About 1903 owing, no doubt, to the fact that the *Temeraire*, Flagship of the Commander-in-Chief, Plymouth Admiral Sir E H Seymour, GCB, OM, was too far up the Hamoaze to be seen from the Sound, the Lords of the Admiralty approved the transfer of the flag of the Port Admiral from the *Temeraire* to the *Impregnable*, the latter then being moored nearer the entrance to the harbour.

Until 1906, she remained moored alone off Cremyll. In that year she was joined by the *Inconstant* and a few years later by the *Black Prince*—the training establishment being ultimately formed of three ships.

Having been replaced by the first class cruiser *Powerful*, the *Impregnable* (ex-*Howe*) paid off on the 1 December 1919. She reverted to her former name *Bulwark* and was finally sold in 1921.

She left Devonport on Thursday, 1 September 1921 in tow of two tugs to be broken up on the Thames by Messrs Garnham and Sons.

HMS **Bulwark***(4) as HMS* **Impregnable** *a training ship for boys at Devonport (1901)*

The Fifth Bulwark
1899 to 1914

A London Class twin screw armoured battleship, designed by Sir William Henry White, KCB, Director of Naval Construction and Controller of the Navy, was laid down in Devonport Dockyard on 20 March 1899. She was launched within 7 months, on 18 October, after being named *Bulwark* by Lady Fairfax, wife of the Commander-in-Chief, Plymouth, Admiral Sir Henry Fairfax, KCB.

The principal technical details were:-

Displacement 15,000 tons; length 400ft(pp) 431¾ft(oa); beam 75ft; draught 26ft 9ins.

Armament four 12″; twelve 6″; sixteen 12 pdrs; and six 3 pdrs.

Torpedo tubes four 18″ (submerged).

Machinery by Messrs R & W Hawthorn, Leslie and Co, of Newcastle-on-Tyne, produced a speed of 18 knots.

Bulwark first commissioned at Devonport on 11 March 1902 for service in the Mediterranean as Flagship of the Commander-in-Chief, Admiral Sir Compton E Domvile KCB. She recommissioned at Devonport on 3 January 1905 for the same service, wearing the flag of Admiral Lord Charles Beresford, GCVO, KCB. Returning to Devonport on 31 January, 1907, *Bulwark* paid off on 11 February but recommissioned the next day with a nucleus crew from the *Resolution*, but this time for service as Flagship of the Rear Admiral in Command of the Home Fleet at the Nore.

On 26 October 1907 during the passage of the Nore Division of the Home Fleet from Queensferry to Sheerness on conclusion of exercises, the *Bulwark* touched a sandbank. It happened about 10 miles south east of the Leman Light in the North Sea, whilst the Fleet were altering course to avoid some fishing boats. The damage necessitated docking for the removal of 9 plates between the keel and the fore part of the bilge keel on the starboard side.

On 30 May 1908 Captain R F Scott, the famous Antarctic explorer, joined *Bulwark*—at the time the most junior Captain in command of a battleship.

Paid off at Chatham on 17 August, 1908, she recommissioned the next day with a nucleus crew from the *Majestic*. The crew was completed at Devonport on 3 October for service in the Channel Fleet attached to Devonport.

In 1910 *Bulwark* became Flagship of the Vice Admiral Commanding the Third and Fourth Divisions of the Home Fleet but on 25 April 1911 recommissioned at Chatham for the Third Division of the Home Fleet.

Bulwark paid off at Chatham on 1 September 1911 for refit. During post-refit trials in May, 1912 she twice grounded on the Barrow Deep off the Nore.

At the outbreak of war, she was commanded by Captain G I Sclater, as a unit of the Fifth Battle Squadron in the Channel Fleet. On 26 November 1914 she was lying in Ketthole Reach in the Medway, embarking ammunition, when she was destroyed by an internal explosion. Only 12 of the ship's company were saved.

The Court of Inquiry held in December 1914 reported that the explosion was due to the accidental ignition of ammunition and there was no evidence to support any suggestion of treachery.

HMS **Bulwark***(5) at anchor in Plymouth Sound.*

And The Sixth...

The Royal Navy's Sixth Bulwark...

Affectionately known throughout the Fleet as 'The Rusty B'—A title dubbed upon the ship by her ships company. It was their own adaptation of her official motto
UNDER THY WINGS I WILL (T)RUST

Technical Details—as built

Length overall 737ft 9ins (224.8 metres)
Length at average waterline 691ft (211 metres)
Length between perpendiculars 650ft (198 metres)
Beam on the waterline 90ft (27.4 metres)
Extreme breadth 121ft 1in (36.9 metres)

Displacement 22,000 tons (standard), 27,000 tons (full load).
Draught
Light condition 21′ 3¼″ (6.48m) forward, 22′ 9½″ (6.93m) aft
Loaded condition 27′ 1⅜″ (8.23m) forward, 28′ 1⅛″ (8.53m) aft

Armament 26 x 40mm AA (two 6 barrelled, five twin, four single)
4 x 3pdr saluting guns
Aircraft 45

Machinery. Parsons geared turbines, 2 shafts
Boilers. Four Admiralty 3-drum type.
SHP 78,000
Speed 28 knots

Complement 1,037; c 1390 with Squadrons embarked.

Interim form of angled flight deck (5½°)

6 Arrester wires

Cost £10,386,000 (excluding guns, aircraft and equipment)

Deck recognition letter B

Pennant Number R08

Acknowledgements

The authors are most grateful to the former Commanding Officers of HMS *Bulwark* for the loan of their photograph albums which have been invaluable in the preparation of this book.

The help offered by the Ministry of Defence, Fleet Air Arm Museum, Fleet Photographic Unit, RM Museum and Lt Cdr B. Warlow RN is gratefully acknowledged.

They are too numerous to mention, but former officers and men of HMS *Bulwark*'s ships company have also helped with the loan of material which has also been used within.

Finis

And, as the flag comes down with the setting sun,
    Final seal on that which we'd begun,
    A silent cheer will explode and rise,
From two thousand throats, and rock the skies.
    But mingling there with that throbbing host,
    A nostalgic sigh, like the ghost of a ghost,
    Will linger, long after the rejoicings are past,
To ensure that the memories of *Bulwark* will last.

(From the book of the 1960-61 commission)

Details of the HMS Bulwark Association are available from the Hon Secretary at 15, Ridgeway Close, Paulsgrove, Portsmouth PO6 4LT.